Clairvoyance

Charles Webster Leadbeater

IAP © 2009

Printed in Scotts Valley, CA – USA.

Leadbeater, Charles Webster.

Clairvoyance / Charles Webster Leadbeater – 1st ed.

1. **Body, Mind & Spirit**

Book Cover: © IAP

CONTENTS

CHAPTER 1

WHAT CLAIRVOYANCE IS

Clairvoyance means literally nothing more than "clear-seeing," and it is a word which has been sorely misused, and even degraded so far as to be employed to describe the trickery of a mountebank in a variety show. Even in its more restricted sense it covers a wide range of phenomena, differing so greatly in character that it is not easy to give a definition of the word which shall be at once succinct and accurate. It has been called "spiritual vision," but no rendering could well be more misleading than that, for in the vast majority of cases there is no faculty connected with it which has the slightest claim to be honored by so lofty a name.

For the purpose of this treatise we may, perhaps, define it as the power to see what is hidden from ordinary physical sight. It will be as well to premise that it is very frequently (though by no means always) accompanied by what is called clairaudience, or the power to hear what would be inaudible to the ordinary physical ear; and we will for the nonce take our title as covering this faculty also, in order to avoid the clumsiness of perpetually using two long words where one will suffice.

Let me make two points clear before I begin. First, I am not writing for those who do not believe that there is such a thing as clairvoyance, nor am I seeking to convince those who are in doubt about the matter. In so small a work as this I have no space for that; such people must study the many books containing lists of cases, or make experiments for themselves along mesmeric lines. I am addressing myself to the better-instructed class who know that clairvoyance exists, and are sufficiently interested in the subject to be glad of information as to its methods and possibilities; and I would assure them that what I write is the result of much careful study and experiment, and that though some of the powers which I shall have to describe may seem new and wonderful to them, I mention no single one of which I have not myself seen examples.

Secondly, though I shall endeavor to avoid technicalities as far as possible, yet as I am writing in the main for students of Theosophy, I shall feel myself at liberty sometimes to use, for brevity's sake and without detailed explanation, the ordinary Theosophical terms with which I may safely assume them to be familiar.

Should this little book fall into the hands of any to whom the occasional use of such terms constitutes a difficulty, I can only apologize to them and refer them for these preliminary explanations to any elementary Theosophical work, such as Mrs. Besant's *Ancient Wisdom* or *Man and His Bodies*. The truth is that the whole Theosophical system hangs together so closely, and its various parts are so interdependent, that to give a full explanation of every term used would necessitate an exhaustive treatise on Theosophy as a preface even to this short account of clairvoyance.

Before a detailed explanation of clairvoyance can usefully be attempted, however, it will be necessary for us to devote a little time to some preliminary considerations, in order that we may have clearly in mind a few broad facts as to the different planes on which clairvoyant vision may be exercised, and the conditions which render its exercise possible.

We are constantly assured in Theosophical literature that all these higher faculties are presently to be the heritage of mankind in general--that the capacity of clairvoyance, for example, lies latent in every one, and that those in whom it already manifests itself are simply in that one particular a little in advance of the rest of us. Now this statement is a true one, and yet it seems quite vague and unreal to the majority of people, simply because they regard such a faculty as something absolutely different from anything they have yet experienced, and feel fairly confident that they themselves, at any rate, are not within measurable distance of its development.

It may help to dispel this sense of unreality if we try to understand that clairvoyance, like so many other things in nature, is mainly a question of vibrations, and is in fact nothing but an extension of powers which we are all using every day of our lives. We are living all the while surrounded by a vast sea of mingled air and ether, the latter inter-penetrating the former, as it does all physical matter; and it is chiefly by means of vibrations in that vast sea of matter that impressions reach us from the outside. This much we all know, but it may perhaps never have occurred to many of us that the number of these vibrations to which we are capable of responding is in reality quite infinitesimal.

Up among the exceedingly rapid vibrations which affect the ether there is a certain small section--a *very* small section--to which the retina of the human eye is capable of responding, and these particular vibrations produce in us the sensation which we call light. That is to say, we are capable of seeing only those objects from which light of that particular kind can either issue or be reflected.

In exactly the same way the tympanum of the human ear is capable of responding to a certain very small range of comparatively slow vibrations--slow enough to affect the air which surrounds us; and so the only sounds which we can hear are those made by objects which are able to vibrate at some rate within that particular range.

In both cases it is a matter perfectly well known to science that there are large numbers of vibrations both above and below these two sections, and that consequently there is much light that we cannot see, and there are many sounds to which our ears are deaf. In the case of light the action of these higher and lower vibrations is easily perceptible in the effects produced by the actinic rays at one end of the spectrum and the heat rays at the other.

As a matter of fact there exist vibrations of every conceivable degree of rapidity, filling the whole vast space intervening between the slow sound waves and the swift light waves; nor is even that all, for there are undoubtedly vibrations slower than those of sound, and a whole infinity of them which are swifter than those known to us as light. So we begin to understand that the vibrations by which we see and hear are only like two tiny groups of a few

7

strings selected from an enormous harp of practically infinite extent, and when we think how much we have been able to learn and infer from the use of those minute fragments, we see vaguely what possibilities might lie before us if we were enabled to utilize the vast and wonderful whole.

Another fact which needs to be considered in this connection is that different human beings vary considerably, though within relatively narrow limits, in their capacity of response even to the very few vibrations which are within reach of our physical senses. I am not referring to the keenness of sight or of hearing that enables one man to see a fainter object or hear a slighter sound than another; it is not in the least a question of strength of vision, but of extent of susceptibility.

For example, if anyone will take a good bisulphide of carbon prism, and by its means throw a clear spectrum on a sheet of white paper, and then get a number of people to mark upon the paper the extreme limits of the spectrum as it appears to them, he is fairly certain to find that their powers of vision differ appreciably. Some will see the violet extending much farther than the majority do; others will perhaps see rather less violet than most, while gaining a corresponding extension of vision at the red end. Some few there will perhaps be who can see farther than ordinary at both ends, and these will almost certainly be what we call sensitive people--susceptible in fact to a greater range of vibrations than are most men of the present day.

In hearing, the same difference can be tested by taking some sound which is just not too high to be audible--on the very verge of audibility as it were--and discovering how many among a given number of people are able to hear it. The squeak of a bat is a familiar instance of such a sound, and experiment will show that on a summer evening, when the whole air is full of the shrill, needle-like cries of these little animals, quite a large number of men will be absolutely unconscious of them, and unable to hear anything at all.

Now these examples clearly show that there is no hard-and-fast limit to man's power of response to either etheric or aerial vibrations, but that some among us already have that power to a wider extent than others; and it will even be found that the same man's capacity varies on different occasions. It is therefore not difficult for us to imagine that it might be possible for a man to develop this power, and thus in time to learn to see much that is invisible to his fellow-men, and hear much that is inaudible to them, since we know perfectly well that enormous numbers of these additional vibrations do exist, and are simply, as it were, awaiting recognition.

The experiments with the Röntgen rays give us an example of the startling results which are produced when even a very few of these additional vibrations are brought within human ken, and the transparency to these rays of many substances hitherto considered opaque at once shows us one way at least in which we may explain such elementary clairvoyance as is involved in reading a letter inside a closed box, or describing those present in an adjoining apartment. To learn to see by means of the Röntgen rays in addition to those ordinarily employed would be quite sufficient to enable anyone to perform a feat of magic of this order.

So far we have thought only of an extension of the purely physical senses of man; and when we remember that a man's etheric body is in reality merely the finer part of his physical frame, and that therefore all his sense organs contain a large amount of etheric matter of various degrees of density, the capacities of which are still practically latent in most of us, we shall see that even if we confine ourselves to this line of development alone there are enormous possibilities of all kinds already opening out before us.

But besides and beyond all this we know that man possesses an astral and a mental body, each of which can in process of time be aroused into activity, and will respond in turn to the vibrations of the matter of its own plane, thus opening up before the Ego, as he learns to function through these vehicles, two entirely new and far wider worlds of knowledge and power. Now these new worlds, though they are all around us and freely inter-penetrate one another, are not to be thought of as distinct and entirely unconnected in substance, but rather as melting the one into the other, the lowest astral forming a direct series with the highest physical, just as the lowest mental in its turn forms a direct series with the highest astral. We are not called upon in thinking of them

9

to imagine some new and strange kind of matter, but simply to think of the ordinary physical kind as subdivided so very much more finely and vibrating so very much more rapidly as to introduce us to what are practically entirely new conditions and qualities.

It is not then difficult for us to grasp the possibility of a steady and progressive extension of our senses, so that both by sight and by hearing we may be able to appreciate vibrations far higher and far lower than those which are ordinarily recognized. A large section of these additional vibrations will still belong to the physical plane, and will merely enable us to obtain impressions from the etheric part of that plane, which is at present as a closed book to us. Such impressions will still be received through the retina of the eye; of course they will affect its etheric rather than its solid matter, but we may nevertheless regard them as still appealing only to an organ specialized to receive them, and not to the whole surface of the etheric body.

There are some abnormal cases, however, in which other parts of the etheric body respond to these additional vibrations as readily as, or even more readily than, the eye. Such vagaries are explicable in various ways, but principally as effects of some partial astral development, for it will be found that the sensitive parts of the body almost invariably correspond with one or other of the *chakrams*, or centers of vitality in the astral body. And though, if astral consciousness be not yet developed, these centers may not be available on their own plane, they are still strong enough to stimulate into keener activity the etheric matter which they inter-penetrate.

When we come to deal with the astral senses themselves the methods of working are very different. The astral body has no specialized sense-organs--a fact which perhaps needs some explanation, since many students who are trying to comprehend its physiology seem to find it difficult to reconcile with the statements that have been made as to the perfect inter-penetration of the physical body by astral matter, the exact correspondence between the two vehicles, and the fact that every physical object has necessarily its astral counterpart.

Now all these statements are true, and yet it is quite possible for people who do not normally see astrally to misunderstand them. Every order of physical matter has its corresponding order of astral matter in constant association with it--not to be separated from it except by a very considerable exertion of occult force, and even then only to be held apart from it as long as force is being definitely exerted to that end. But for all that the relation of the astral particles one to another is far looser than is the case with their physical correspondences.

In a bar of iron, for example, we have a mass of physical molecules in the solid condition--that is to say, capable of comparatively little change in their relative positions, though each vibrating with immense rapidity in its own sphere. The astral counterpart of this consists of what we often call solid astral matter--that is, matter of the lowest and densest sub-plane of the astral; but nevertheless its particles are constantly and rapidly changing their relative position, moving among one another as easily as those of a liquid on the physical plane might do. So that there is no permanent association between any one physical particle and that amount of astral matter which happens at any given moment to be acting as its counterpart.

This is equally true with respect to the astral body of man, which for our purpose at the moment we may regard as consisting of two parts--the denser aggregation which occupies the exact position of the physical body, and the cloud of rarer astral matter which surrounds that aggregation. In both these parts, and between them both, there is going on at every moment of time the rapid inter-circulation of the particles which has been described, so that as one watches the movement of the molecules in the astral body one is reminded of the appearance of those in fiercely boiling water.

This being so, it will be readily understood that though any given organ of the physical body must always have as its counterpart a certain amount of astral matter, it does not retain the same particles for more than a few seconds at a time, and consequently there is nothing corresponding to the specialization of physical nerve-matter into optic or auditory nerves, and so on. So that though the physical eye or ear has undoubtedly always its counterpart of astral matter, that particular fragment of astral

11

matter is no more (and no less) capable of responding to the vibrations which produce astral sight or astral hearing than any other part of the vehicle.

It must never be forgotten that though we constantly have to speak of "astral sight" or "astral hearing" in order to make ourselves intelligible, all that we mean by those expressions is the faculty of responding to such vibrations as convey to the man's consciousness, when he is functioning in his astral body, information of the same character as that conveyed to him by his eyes and ears while he is in the physical body. But in the entirely different astral conditions, specialized organs are not necessary for the attainment of this result; there is matter in every part of the astral body which is capable of such response, and consequently the man functioning in that vehicle sees equally well objects behind him, beneath him, above him, without needing to turn his head.

There is, however, another point which it would hardly be fair to leave entirely out of account, and that is the question of the *chakrams* referred to above. Theosophical students are familiar with the idea of the existence in both the astral and the etheric bodies of man of certain centers of force which have to be vivified in turn by the sacred serpent-fire as the man advances in evolution. Though these cannot be described as organs in the ordinary sense of the word, since it is not through them that the man sees or hears, as he does in physical life through eyes and ears, yet it is apparently very largely upon their vivification that the power of exercising these astral senses depends, each of them as it is developed giving to the whole astral body the power of response to a new set of vibrations.

Neither have these centers, however, any permanent collection of astral matter connected with them. They are simply vortices in the matter of the body--vortices through which all the particles pass in turn--points, perhaps, at which the higher force from planes above impinges upon the astral body. Even this description gives but a very partial idea of their appearance, for they are in reality four-dimensional vortices, so that the force which comes through them and is the cause of their existence seems to well up from nowhere. But at any rate, since all particles in turn pass through each of them, it will be clear that it is thus possible for each in turn to evoke in all the particles of the body the power of receptivity to a certain set of vibrations, so that all the astral senses are equally active in all parts of the body.

The vision of the mental plane is again totally different, for in this case we can no longer speak of separate senses such as sight and hearing, but rather have to postulate one general sense which responds so fully to the vibrations reaching it that when any object comes within its cognition it at once comprehends it fully, and as it were sees it, hears it, feels it, and knows all there is to know about it by the one instantaneous operation. Yet even this wonderful faculty differs in degree only and not in kind from those which are at our command at the present time; on the mental plane, just as on the physical, impressions are still conveyed by means of vibrations travelling from the object seen to the seer.

On the buddhic plane we meet for the first time with a quite new faculty having nothing in common with those of which we have spoken, for there a man cognizes any object by an entirely different method, in which external vibrations play no part. The object becomes part of himself, and he studies it from the inside instead of from the outside. But with *this* power ordinary clairvoyance has nothing to do.

The development, either entire or partial, of any one of these faculties would come under our definition of clairvoyance--the power to see what is hidden from ordinary physical sight. But these faculties may be developed in various ways, and it will be well to say a few words as to these different lines.

We may presume that if it were possible for a man to be isolated during his evolution from all but the gentlest outside influences, and to unfold from the beginning in perfectly regular and normal fashion, he would probably develop his senses in regular order also. He would find his physical senses gradually extending their scope until they responded to all the physical vibrations, of etheric as well as of denser matter; then in orderly sequence would come sensibility to the coarser part of the astral plane, and presently the finer part also would be included, until in due course the faculty of the mental plane dawned in its turn.

In real life, however, development so regular as this is hardly ever known, and many a man has occasional flashes of astral consciousness without any awakening of etheric vision at all. And this irregularity of development is one of the principal causes of man's extraordinary liability to error in matters of clairvoyance--a liability from which there is no escape except by a long course of careful training under a qualified teacher.

Students of Theosophical literature are well aware that there are such teachers to be found--that even in this materialistic nineteenth century the old saying is still true, that "when the pupil is ready, the Master is ready also," and that "in the hall of learning, when he is capable of entering there, the disciple will always find his Master." They are well aware also that only under such guidance can a man develop his latent powers in safety and with certainty, since they know how fatally easy it is for the untrained clairvoyant to deceive himself as to the meaning and value of what he sees, or even absolutely to distort his vision completely in bringing it down into his physical consciousness.

It does not follow that even the pupil who is receiving regular instruction in the use of occult powers will find them unfolding themselves exactly in the regular order which was suggested above as probably ideal. His previous progress may not have been such as to make this for him the easiest or most desirable road; but at any rate he is in the hands of one who is perfectly competent to be his guide in spiritual development, and he rests in perfect contentment that the way along which he is taken will be that which is the best way for him.

Another great advantage which he gains is that whatever faculties he may acquire are definitely under his command and can be used fully and constantly when he needs them for his Theosophical work; whereas in the case of the untrained man such powers often manifest themselves only very partially and spasmodically, and appear to come and go, as it were, at their own sweet will.

It may reasonably be objected that if clairvoyant faculty is, as stated, a part of the occult development of man, and so a sign of a certain amount of progress along that line, it seems strange that it should often be possessed by primitive peoples, or by the ignorant and uncultured among our own race--persons who are obviously quite undeveloped, from whatever point of view one regards them. No doubt this does appear remarkable at first sight but the fact is that the sensitiveness of the savage or of the coarse and vulgar European ignoramus is not really at all the same thing as the faculty of his properly trained brother, nor is it arrived at in the same way.

An exact and detailed explanation of the difference would lead us into rather recondite technicalities, but perhaps the general idea of the distinction between the two may be caught from an example taken from the very lowest plane of clairvoyance, in close contact with the denser physical. The etheric double in man is in exceedingly close relation to his nervous system, and any kind of action upon one of them speedily reacts on the other. Now in the sporadic appearance of etheric sight in the savage, whether of Central Africa or of Western Europe, it has been observed that the corresponding nervous disturbance is almost entirely in the sympathetic system, and that the whole affair is practically beyond the man's control--is in fact a sort of massive sensation vaguely belonging to the whole etheric body, rather than an exact and definite sense-perception communicated through a specialized organ.

As in later races and amid higher development the strength of the man is more and more thrown into the evolution of the mental faculties, this vague sensitiveness usually disappears; but still later, when the spiritual man begins to unfold, he regains his clairvoyant power. This time, however, the faculty is a precise and exact one, under the control of the man's will, and exercised through a definite sense-organ; and it is noteworthy that any nervous action set up in sympathy with it is now almost exclusively in the cerebro-spinal system.

On this subject Mrs. Besant writes:--"The lower forms of psychism are more frequent in animals and in very unintelligent human beings than in men and women in whom the intellectual powers are well developed. They appear to be connected with the sympathetic system, not with the cerebro-spinal. The large nucleated ganglionic cells in this system contain a very large proportion of etheric matter, and are hence more easily affected by the coarser astral vibrations than are the cells in which the proportion is less. As the cerebro-spinal system develops, and the brain becomes more highly evolved, the sympathetic system subsides into a subordinate position, and the sensitiveness to psychic vibrations is dominated by the stronger and more active vibrations of the higher nervous system. It is true that at a later stage of evolution psychic sensitiveness reappears, but it is then developed in connection with the cerebro-spinal centers, and is brought under the control of the will. But the hysterical and ill-regulated psychism of which we see so many lamentable examples is due to the small development of the brain and the dominance of the sympathetic system."

Occasional flashes of clairvoyance do, however, sometimes come to the highly cultured and spiritual-minded man, even though he may never have heard of the possibility of training such a faculty. In his case such glimpses usually signify that he is approaching that stage in his evolution when these powers will naturally begin to manifest themselves, and their appearance should serve as an additional stimulus to him to strive to maintain that high standard of moral purity and mental balance without which clairvoyance is a curse and not a blessing to its possessor.

Between those who are entirely unimpressible and those who are in full possession of clairvoyant power there are many intermediate stages. One to which it will be worth while to give a passing glance is the stage in which a man, though he has no clairvoyant faculty in ordinary life, yet exhibits it more or less fully under the influence of mesmerism. This is a case in which the psychic nature is already sensitive, but the consciousness is not yet capable of functioning in it amidst the manifold distractions of physical life. It needs to be set free by the temporary suspension of the outer senses in the mesmeric trance before it can use the diviner faculties which are but just beginning to dawn within it. But of course even in the mesmeric trance there are innumerable degrees of lucidity, from the ordinary patient who is blankly unintelligent to the man whose power of sight is fully under the control of the operator, and can be directed whithersoever he wills, or to the more advanced stage in which, when the consciousness is once set free, it escapes altogether from the grasp of the magnetizer, and soars into fields of exalted vision where it is entirely beyond his reach.

Another step along the same path is that upon which such perfect suppression of the physical as that which occurs in the hypnotic trance is not necessary, but the power of supernormal sight, though still out of reach during waking life, becomes available when the body is held in the bonds of ordinary sleep. At this stage of development stood many of the prophets and seers of whom we read, who were "warned of God in a dream," or communed with beings far higher than themselves in the silent watches of the night.

Most cultured people of the higher races of the world have this development to some extent: that is to say, the senses of their astral bodies are in full working order, and perfectly capable of receiving impressions from objects and entities of their own plane. But to make that fact of any use to them down here in the physical body, two changes are usually necessary; first, that the Ego shall be awakened to the realities of the astral plane, and induced to emerge from the chrysalis formed by his own waking thoughts, and look round him to observe and to learn; and secondly, that the consciousness shall be so far retained during the return of the Ego into his physical body as to enable him to impress upon his physical brain the recollection of what he has seen or learnt.

If the first of these changes has taken place, the second is of little importance, since the Ego, the true man, will be able to profit by the information to be obtained upon that plane, even though he may not have the satisfaction of bringing through any remembrance of it into his waking life down here.

Students often ask how this clairvoyant faculty will first be manifested in themselves--how they may know when they have reached the stage at which its first faint foreshadowings are beginning to be visible. Cases differ so widely that it is impossible to give to this question any answer that will be universally applicable.

Some people begin by a plunge, as it were, and under some unusual stimulus become able just for once to see some striking vision; and very often in such a case, because the experience does not repeat itself, the seer comes in time to believe that on that occasion he must have been the victim of hallucination. Others begin by becoming intermittently conscious of the brilliant colors and vibrations of the human aura; yet others find themselves with increasing frequency seeing and hearing something to which those around them are blind and deaf; others, again, see faces, landscapes, or colored clouds floating before their eyes in the dark before they sink to rest; while perhaps the commonest experience of all is that of those who begin to recollect with greater and greater clearness what they have seen and heard on the other planes during sleep.

Having now to some extent cleared our ground, we may proceed to consider the various phenomena of clairvoyance.

They differ so widely both in character and in degree that it is not very easy to decide how they can most satisfactorily be classified. We might, for example, arrange them according to the kind of sight employed--whether it were mental, astral, or merely etheric. We might divide them according to the capacity of the clairvoyant, taking into consideration whether he was trained or untrained; whether his vision was regular and under his command, or spasmodic and independent of his volition; whether he could exercise it only when under mesmeric influence, or whether that assistance was unnecessary for him; whether he was able to use his faculty when awake in the physical body, or whether it was available only when he was temporarily away from that body in sleep or trance.

All these distinctions are of importance, and we shall have to take them all into consideration as we go on, but perhaps on the whole the most useful classification will be one something on the lines of that adopted by Mr. Sinnett in his *Rationale of Mesmerism* a book, by the way, which all students of clairvoyance ought to read. In dealing with the phenomena, then, we will arrange them rather according to the capacity of the sight employed than to the plane upon which it is exercised, so that we may group instances of clairvoyance under some such headings as these:

1. Simple clairvoyance--that is to say, a mere opening of sight,

enabling its possessor to see whatever astral or etheric entities happen to be present around him, but not including the power of observing either distant places or scenes belonging to any other time than the present.

2. Clairvoyance in space--the capacity to see scenes or events removed from the seer in space, and either too far distant for ordinary observation or concealed by intermediate objects.

3. Clairvoyance in time--that is to say, the capacity to see objects or events which are removed from the seer in time, or, in other words, the power of looking into the past or the future.

CHAPTER 2

SIMPLE CLAIRVOYANCE: FULL

We have defined this as a mere opening of etheric or astral sight, which enables the possessor to see whatever may be present around him on corresponding levels, but is not usually accompanied by the power of seeing anything at a great distance or of reading either the past or the future. It is hardly possible altogether to exclude these latter faculties, for astral sight necessarily has considerably greater extension than physical, and fragmentary pictures of both past and future are often casually visible even to clairvoyants who do not know how to seek specially for them; but there is nevertheless a very real distinction between such incidental glimpses and the definite power of projection of the sight either in space or time.

We find among sensitive people all degrees of this kind of clairvoyance, from that of the man who gets a vague impression which hardly deserves the name of sight at all, up to the full possession of etheric and astral vision respectively. Perhaps the simplest method will be for us to begin by describing what would be visible in the case of this fuller development of the power, as the cases of its partial possession will then be seen to fall naturally into their places.

Let us take the etheric vision first. This consists simply, as has already been said, in susceptibility to a far larger series of physical vibrations than ordinary, but nevertheless its possession brings into view a good deal to which the majority of the human race still remains blind. Let us consider what changes its acquisition produces in the aspect of familiar objects, animate and inanimate, and then see to what entirely new factors it introduces us. But it must be remembered that what I am about to describe is the result of the full and perfectly-controlled possession of the faculty only, and that most of the instances met with in real life will be likely to fall far short of it in one direction or another.

The most striking change produced in the appearance of inanimate objects by the acquisition of this faculty is that most of them become almost transparent, owing to the difference in wave-length of some of the vibrations to which the man has now become susceptible. He finds himself capable of performing with the utmost ease the proverbial feat of "seeing through a brick wall," for to his newly-acquired vision the brick wall seems to have a consistency no greater than that of a light mist. He therefore sees what is going on in an adjoining room almost as though no intervening wall existed; he can describe with accuracy the contents of a locked box, or read a sealed letter; with a little practice he can find a given passage in a closed book. This last feat, though perfectly easy to astral vision, presents considerable difficulty to one using etheric sight, because of the fact that each page has to be looked at *through* all those which happen to be superimposed upon it.

It is often asked whether under these circumstances a man sees always with this abnormal sight, or only when he wishes to do so. The answer is that if the faculty is perfectly developed it will be entirely under his control, and he can use that or his more ordinary vision at will. He changes from one to the other as readily and naturally as we now change the focus of our eyes when we look up from our book to follow the motions of some object a mile away. It is, as it were, a focusing of consciousness on the one or the other aspect of what is seen; and though the man would have quite clearly in his view the aspect upon which his attention was for the moment fixed, he would always be vaguely conscious of the other aspect too, just as when we focus our sight upon any object held in our hands we yet vaguely see the opposite wall of the room as a background.

Another curious change, which comes from the possession of this sight, is that the solid ground upon which the man walks becomes to a certain extent transparent to him, so that he is able to see down into it to a considerable depth, much as we can now see into fairly clear water. This enables him to watch a creature burrowing underground, to distinguish a vein of coal or of metal if not too far below the surface, and so on.

The limit of etheric sight when looking through solid matter appears to be analogous to that imposed upon us when looking through water or mist. We cannot see beyond a certain distance, because the medium through which we are looking is not perfectly transparent.

The appearance of animate objects is also considerably altered for the man who has increased his visual powers to this extent. The bodies of men and animals are for him in the main transparent, so that he can watch the action of the various internal organs, and to some extent diagnose some of their diseases.

The extended sight also enables him to perceive, more or less clearly, various classes of creatures, elemental and otherwise, whose bodies are not capable of reflecting any of the rays within the limit of the spectrum as ordinarily seen. Among the entities so seen will be some of the lower orders of nature-spirits--those whose bodies are composed of the denser etheric matter. To this class belong nearly all the fairies, gnomes, and brownies, about whom there are still so many stories remaining among Scotch and Irish mountains and in remote country places all over the world.

The vast kingdom of nature-spirits is in the main an astral kingdom, but still there is a large section of it which appertains to the etheric part of the physical plane, and this section, of course, is much more likely to come within the ken of ordinary people than the others. Indeed, in reading the common fairy stories one frequently comes across distinct indications that it is with this class that we are dealing. Any student of fairy lore will remember how often mention is made of some mysterious ointment or drug, which when applied to a man's eyes enables him to see the members of the fairy commonwealth whenever he happens to meet them.

The story of such an application and its results occurs so constantly and comes from so many different parts of the world that there must certainly be some truth behind it, as there always is behind really universal popular tradition. Now no such anointing of the eyes alone could by any possibility open a man's astral vision, though certain ointments rubbed over the whole body will very greatly assist the astral body to leave the physical in full consciousness--a fact the knowledge of which seems to have survived even to medieval times, as will be seen from the evidence given at some of the trials for witchcraft. But the application to the physical eye might very easily so stimulate its sensitiveness as to make it susceptible to some of the etheric vibrations.

The story frequently goes on to relate how when the human being who has used this mystical ointment betrays his extended vision in some way to a fairy, the latter strikes or stabs him in the eye, thus depriving him not only of the etheric sight, but of that of the denser physical plane as well. (See *The Science of Fairy Tales*, by E. S. Hartland, in the "Contemporary Science" series--or indeed almost any extensive collection of fairy stories.) If the sight acquired had been astral, such a proceeding would have been entirely unavailing, for no injury to the physical apparatus would affect an astral faculty; but if the vision produced by the ointment were etheric, the destruction of the physical eye would in most cases at once extinguish it, since that is the mechanism by means of which it works.

Anyone possessing this sight of which we are speaking would also be able to perceive the etheric double of man; but since this is so nearly identical in size with the physical, it would hardly be likely to attract his attention unless it were partially projected in trance or under the influence of anesthetics. After death, when it withdraws entirely from the dense body, it would be clearly visible to him, and he would frequently see it hovering over newly made graves as he passed through a churchyard or cemetery. If he were to attend a spiritualistic séance he would see the etheric matter oozing out from the side of the medium, and could observe the various ways in which the communicating entities make use of it.

Another fact which could hardly fail soon to thrust itself upon his notice would be the extension of his perception of color. He would find himself able to see several entirely new colors, not in the least resembling any of those included in the spectrum as we at present know it, and therefore of course quite indescribable in any terms at our command. And not only would he see new objects that were wholly of these new colors, but he would also discover that modifications had been introduced into the color of many objects with which he was quite familiar, according to whether they had or had not some tinge of these new hues intermingled with the old. So that two surfaces of color which to ordinary eyes appeared to match perfectly would often present distinctly different shades to his keener sight.

We have now touched upon some of the principal changes which would be introduced into a man's world when he gained etheric sight; and it must always be remembered that in most cases a corresponding change would at the same time be brought about in his other senses also, so that he would be capable of hearing, and perhaps even of feeling, more than most of those around him. Now supposing that in addition to this he obtained the sight of the astral plane, what further changes would be observable?

Well, the changes would be many and great; in fact, a whole new world would open before his eyes. Let us consider its wonders briefly in the same order as before, and see first what difference there would be in the appearance of inanimate objects. On this point I may begin by quoting a recent quaint answer given in *The Vâhan.*

"There is a distinct difference between etheric sight and astral sight, and it is the latter which seems to correspond to the fourth dimension.

"The easiest way to understand the difference is to take an example. If you looked at a man with both the sights in turn, you would see the buttons at the back of his coat in both cases; only if you used etheric sight you would see them *through* him, and would see the shank-side as nearest to you, but if you looked astrally, you would see it not only like that, but just as if you were standing behind the man as well.

"Or if you were looking etherically at a wooden cube with writing on all its sides, it would be as though the cube were glass, so that you could see through it, and you would see the writing on the opposite side all backwards, while that on the right and left sides would not be clear to you at all unless you moved, because you would see it edgewise. But if you looked at it astrally you would see all the sides at once, and all the right way up, as though the whole cube had been flattened out before you, and you would see every particle of the inside as well--not *through* the others, but all flattened out. You would be looking at it from another direction, at right angles to all the directions that we know.

"If you look at the back of a watch etherically you see all the wheels through it, and the face *through them*, but backwards; if you look at it astrally, you see the face right way up and all the wheels lying separately, but nothing on the top of anything else."

Here we have at once the keynote, the principal factor of the change; the man is looking at everything from an absolutely new point of view, entirely outside of anything that he has ever imagined before. He has no longer the slightest difficulty in reading any page in a closed book, because he is not now looking at it through all the other pages before it or behind it, but is looking straight down upon it as though it were the only page to be seen. The depth at which a vein of metal or of coal may lie is no longer a barrier to his sight of it, because he is not now looking through the intervening depth of earth at all. The thickness of a wall, or the number of walls intervening between the observer and the object, would make a great deal of difference to the clearness of the etheric sight; they would make no difference whatever to the astral sight, because on the astral plane they would *not* intervene between the observer and the object. Of course that sounds paradoxical and impossible, and it *is* quite inexplicable to a mind not specially trained to grasp the idea; yet it is none the less absolutely true.

This carries us straight into the middle of the much-vexed question of the fourth dimension--a question of the deepest interest, though one that we cannot pretend to discuss in the space at our disposal. Those who wish to study it as it deserves are recommended to begin with Mr. C. H. Hinton's *Scientific Romances* or Dr. A. T. Schofield's *Another World,* and then follow on with the former author's larger work, *A New Era of Thought.* Mr. Hinton not only claims to be able himself to grasp mentally some of the simpler fourth-dimensional figures, but also states that anyone who will take the trouble to follow out his directions may with perseverance acquire that mental grasp likewise. I am not certain that the power to do this is within the reach of everyone, as he thinks, for it appears to me to require considerable mathematical ability; but I can at any rate bear witness that the tesseract or fourth-dimensional cube which he describes is a reality, for it is quite a familiar figure upon the astral plane. He has now perfected a new method of representing the several dimensions by colors instead of by arbitrary written symbols. He states that this will very much simplify the study, as the reader will be able to distinguish instantly by sight any part or feature of the tesseract. A full description of this new method, with plates, is said to be ready for the press, and is expected to appear within a year, so that intending students of this fascinating subject might do well to await its publication.

I know that Madame Blavatsky, in alluding to the theory of the fourth dimension, has expressed an opinion that it is only a clumsy way of stating the idea of the entire permeability of matter, and that Mr. W. T. Stead has followed along the same lines, presenting the conception to his readers under the name of *throughth.* Careful, oft-repeated and detailed investigation does, however, seem to show quite conclusively that this explanation does not cover all the facts. It is a perfect description of etheric vision, but the further and quite different idea of the fourth dimension as expounded by Mr. Hinton is the only one which gives any kind of explanation down here of the constantly-observed facts of astral vision. I would therefore venture deferentially to suggest that when Madame Blavatsky wrote as she did, she had in mind etheric vision and not astral, and that the extreme applicability of the phrase to this other and higher faculty, of which she was not at the moment thinking, did not occur to her.

The possession of this extraordinary and scarcely expressible power, then, must always be borne in mind through all that follows. It lays every point in the interior of every solid body absolutely open to the gaze of the seer, just as every point in the interior of a circle lies open to the gaze of a man looking down upon it.

But even this is by no means all that it gives to its possessor. He sees not only the inside as well as the outside of every object, but also its astral counterpart. Every atom and molecule of physical matter has its corresponding astral atoms and molecules, and the mass which is built up out of these is clearly visible to our clairvoyant. Usually the astral of any object projects somewhat beyond the physical part of it, and thus metals, stones and other things are seen surrounded by an astral aura.

It will be seen at once that even in the study of inorganic matter a man gains immensely by the acquisition of this vision. Not only does he see the astral part of the object at which he looks, which before was wholly hidden from him; not only does he see much more of its physical constitution than he did before, but even what was visible to him before is now seen much more clearly and truly. A moment's consideration will show that his new vision approximates much more closely to true perception than does physical sight. For example, if he looks astrally at a glass cube, its sides will all appear equal, as we know they really are, whereas on the physical plane he sees the further side in perspective--that is, it appears smaller than the nearer side, which is, of course, a mere allusion due to his physical limitations.

When we come to consider the additional facilities which it offers in the observation of animate objects we see still more clearly the advantages of the astral vision. It exhibits to the clairvoyant the aura of plants and animals, and thus in the case of the latter their desires and emotions, and whatever thoughts they may have, are all plainly shown before his eyes.

But it is in dealing with human beings that he will most appreciate the value of this faculty, for he will often be able to help them far more effectually when he guides himself by the information which it gives him.

He will be able to see the aura as far up as the astral body, and though that leaves all the higher part of a man still hidden from his gaze, he will nevertheless find it possible by careful observation to learn a good deal about the higher part from what is within his reach. His capacity of examining the etheric double will give him considerable advantage in locating and classifying any defects or diseases of the nervous system, while from the appearance of the astral body he will be at once aware of all the emotions, passions, desires and tendencies of the man before him, and even of very many of his thoughts also.

As he looks at a person he will see him surrounded by the luminous mist of the astral aura, flashing with all sorts of brilliant colors, and constantly changing in hue and brilliancy with every variation of the person's thoughts and feelings. He will see this aura flooded with the beautiful rose-color of pure affection, the rich blue of devotional feeling, the hard, dull brown of selfishness, the deep scarlet of anger, the horrible lurid red of sensuality, the livid grey of fear, the black clouds of hatred and malice, or any of the other hundredfold indications so easily to be read in it by a practiced eye; and thus it will be impossible for any persons to conceal from him the real state of their feelings on any subject.

These varied indications of the aura are of themselves a study of very deep interest, but I have no space to deal with them in detail here. A much fuller account of them, together with a large number of colored illustrations, will be found in my work on the subject *Man Visible and Invisible*.

Not only does the astral aura show him the temporary result of the emotion passing through it at the moment, but it also gives him, by the arrangement and proportion of its colors when in a condition of comparative rest, a clue to the general disposition and character of its owner. For the astral body is the expression of as much of the man as can be manifested on that plane, so that from what is seen in it much more which belongs to higher planes may be inferred with considerable certainty.

In this judgment of character our clairvoyant will be much helped by so much of the person's thought as expresses itself on the astral plane, and consequently comes within his purview. The true home of thought is on the mental plane, and all thought first manifests itself there as a vibration of the mind-body. But if it be in any way a selfish thought, or if it be connected in any way with an emotion or a desire, it immediately descends into the astral plane, and takes to itself a visible form of astral matter.

In the case of the majority of men almost all thought would fall under one or other of these heads, so that practically the whole of their personality would lie clearly before our friend's astral vision, since their astral bodies and the thought-forms constantly radiating from them would be to him as an open book in which their characteristics were writ so largely that he who ran might read. Anyone wishing to gain some idea as to *how* the thought-forms present themselves to clairvoyant vision may satisfy themselves to some extent by examining the illustrations accompanying Mrs. Besant's valuable article on the subject in *Lucifer* for September 1896.

We have seen something of the alteration in the appearance of both animate and inanimate objects when viewed by one possessed of full clairvoyant sight as far as the astral plane is concerned; let us now consider what entirely new objects he will see. He will be conscious of a far greater fullness in nature in many directions, but chiefly his attention will be attracted by the living denizens of this new world. No detailed account of them can be attempted within the space at our disposal; for that the reader is referred to No. V. of the *Theosophical Manuals*. Here we can do no more than barely enumerate a few classes only of the vast hosts of astral inhabitants.

He will be impressed by the protean forms of the ceaseless tide of elemental essence, ever swirling around him, menacing often, yet always retiring before a determined effort of the will; he will marvel at the enormous army of entities temporarily called out of this ocean into separate existence by the thoughts and wishes of man, whether good or evil. He will watch the manifold tribes of the nature-spirits at their work or at their play; he will sometimes be able to study with ever-increasing delight the magnificent evolution of some of the lower orders of the glorious kingdom of the devas, which corresponds approximately to the angelic host of Christian terminology.

But perhaps of even keener interest to him than any of these will be the human denizens of the astral world, and he will find them divisible into two great classes--those whom we call the living, and those others, most of them infinitely more alive, whom we so foolishly misname the dead. Among the former he will find here and there one wide awake and fully conscious, perhaps sent to bring him some message, or examining him keenly to see what progress he is making; while the majority of his neighbors, when away from their physical bodies during sleep, will drift idly by, so wrapped up in their own cogitations as to be practically unconscious of what is going on around them.

Among the great host of the recently dead he will find all degrees of consciousness and intelligence, and all shades of character--for death, which seems to our limited vision so absolute a change, in reality alters nothing of the man himself. On the day after his death he is precisely the same man as he was the day before it, with the same disposition, the same qualities, the same virtues and vices, save only that he has cast aside his physical body; but the loss of that no more makes him in any way a different man than would the removal of an overcoat. So among the dead our student will find men intelligent and stupid, kind-hearted and morose, serious and frivolous, spiritually-minded and sensually-minded, just as among the living.

Since he can not only see the dead, but speak with them, he can often be of very great use to them, and give them information and guidance which is of the utmost value to them. Many of them are in a condition of great surprise and perplexity, and sometimes even of acute distress, because they find the facts of the next world so unlike the childish legends which are all that popular religion in the West has to offer with reference to this transcendently important subject; and therefore a man who understands this new world and can explain matters is distinctly a friend in need.

In many other ways a man who fully possesses this faculty may be of use to the living as well as to the dead; but of this side of the subject I have already written in my little book on *Invisible Helpers*. In addition to astral entities he will see astral corpses--shades and shells in all stages of decay; but these need only be just mentioned here, as the reader desiring a further account of them will find it in our third and fifth manuals.

Another wonderful result which the full enjoyment of astral clairvoyance brings to a man is that he has no longer any break in consciousness. When he lies down at night he leaves his physical body to the rest which it requires, while he goes about his business in the far more comfortable astral vehicle. In the morning he returns to and re-enters his physical body, but without any loss of consciousness or memory between the two states, and thus he is able to live, as it were, a double life which yet is one, and to be usefully employed during the whole of it, instead of losing one-third of his existence in blank unconsciousness.

Another strange power of which he may find himself in possession (though its full control belongs rather to the still higher devachanic faculty), is that of magnifying at will the minutest physical or astral particle to any desired size, as though by a microscope--though no microscope ever made or ever likely to be made possesses even a thousandth part of this psychic magnifying power. By its means the hypothetical molecule and atom postulated by science become visible and living realities to the occult student, and on this closer examination he finds them to be much more complex in their structure than the scientific man has yet realized them to be. It also enables him to follow with the closest attention and the most lively interest all kinds of electrical, magnetic, and other etheric action; and when some of the specialists in these branches of science are able to develop the power to see those things whereof they write so facilely, some very wonderful and beautiful revelations may be expected.

This is one of the *siddhis* or powers described in Oriental books as accruing to the man who devotes himself to spiritual development, though the name under which it is there mentioned might not be immediately recognizable. It is referred to as "the power of making oneself large or small at will," and the reason of a description which appears so oddly to reverse the fact is that in reality the method by which this feat is performed is precisely that indicated in these ancient books. It is by the use of temporary visual machinery of inconceivable minuteness that the world of the infinitely little is so clearly seen; and in the same way (or rather in the opposite way) it is by temporarily enormously increasing the size of the machinery used that it becomes possible to increase the breadth of one's view--in the physical sense as well as, let us hope, in the moral--far beyond anything that science has ever dreamt of as possible for man. So that the alteration in size is really in the vehicle of the student's consciousness, and not in anything outside of himself; and the old Oriental book has, after all, put the case more accurately than we.

Psychometry and second-sight *in excelsis* would also be among the faculties which our friend would find at his command; but those will be more fitly dealt with under a later heading, since in almost all their manifestations they involve clairvoyance either in space or in time.

I have now indicated, though only in the roughest outlines, what a trained student, possessed of full astral vision, would see in the immensely wider world to which that vision introduced him; but I have said nothing of the stupendous change in his mental attitude which comes from the experiential certainty as to the existence of the soul, its survival after death, the action of the law of karma, and other points of equally paramount importance. The difference between even the profoundest intellectual conviction and the precise knowledge gained by direct personal experience must be felt in order to be appreciated.

CHAPTER 3

SIMPLE CLAIRVOYANCE:
PARTIAL

The experiences of the untrained clairvoyant--and be it remembered that that class includes all European clairvoyants except a very few--will, however, usually fall very far short of what I have attempted to indicate; they will fall short in many different ways--in degree, in variety, or in permanence, and above all in precision.

Sometimes, for example, a man's clairvoyance will be permanent, but very partial, extending only perhaps to one or two classes of the phenomena observable; he will find himself endowed with some isolated fragment of higher vision, without apparently possessing other powers of sight which ought normally to accompany that fragment, or even to precede it. For example, one of my dearest friends has all his life had the power to see the atomic ether and atomic astral matter, and to recognize their structure, alike in darkness or in light, as inter-penetrating everything else; yet he has only rarely seen entities whose bodies are composed of the much more obvious lower ethers or denser astral matter, and at any rate is certainly not permanently able to see them. He simply finds himself in possession of this special faculty, without any apparent reason to account for it, or any recognizable relation to anything else: and beyond proving to him the existence of these atomic planes and demonstrating their arrangement, it is difficult to see of what particular use it is to him at present. Still, there the thing is, and it is an earnest of greater things to come--of further powers still awaiting development.

There are many similar cases--similar, I mean, not in the possession of that particular form of sight (which is unique in my experience), but in showing the development of some one small part of the full and clear vision of the astral and etheric planes. In nine cases out of ten, however, such partial clairvoyance will at the same time lack precision also--that is to say, there will be a good deal of vague impression and inference about it, instead of the clear-cut definition and certainty of the trained man. Examples of this type are constantly to be found, especially among those who advertise themselves as "test and business clairvoyants."

Then, again, there are those who are only temporarily clairvoyant under certain special conditions. Among these there are various subdivisions, some being able to reproduce the state of clairvoyance at will by again setting up the same conditions, while with others it comes sporadically, without any observable reference to their surroundings, and with yet others the power shows itself only once or twice in the whole course of their lives.

To the first of these subdivisions belong those who are clairvoyant only when in the mesmeric trance--who when not so entranced are incapable of seeing or hearing anything abnormal. These may sometimes reach great heights of knowledge and be exceedingly precise in their indications, but when that is so they are usually undergoing a course of regular training, though for some reason unable as yet to set themselves free from the leaden weight of earthly life without assistance.

In the same class we may put those--chiefly Orientals--who gain some temporary sight only under the influence of certain drugs, or by means of the performance of certain ceremonies. The ceremonialist sometimes hypnotizes himself by his repetitions, and in that condition becomes to some extent clairvoyant; more often he simply reduces himself to a passive condition in which some other entity can obsess him and speak through him. Sometimes, again, his ceremonies are not intended to affect himself at all, but to invoke some astral entity who will give him the required information; but of course that is a case of magic, and not of clairvoyance. Both the drugs and the ceremonies are methods emphatically to be avoided by any one who wishes to approach clairvoyance from the higher side, and use it for his own progress and for the helping of others. The Central African medicine-man or witch-doctor and some of the Tartar Shamans are good examples of the type.

Those to whom a certain amount of clairvoyant power has come occasionally only, and without any reference to their own wish, have often been hysterical or highly nervous persons, with whom the faculty was to a large extent one of the symptoms of a disease. Its appearance showed that the physical vehicle was weakened to such a degree that it no longer presented any obstacle in the way of a certain modicum of etheric or astral vision. An extreme example of this class is the man who drinks himself into delirium tremens, and in the condition of absolute physical ruin and impure psychic excitation brought about by the ravages of that fell disease, is able to see for the time some of the loathsome elemental and other entities which he has drawn round himself by his long course of degraded and bestial indulgence. There are, however, other cases where the power of sight has appeared and disappeared without apparent reference to the state of the physical health; but it seems probable that even in those, if they could have been observed closely enough, some alteration in the condition of the etheric double would have been noticed.

Those who have only one instance of clairvoyance to report in the whole of their lives are a difficult band to classify at all exhaustively, because of the great variety of the contributory circumstances. There are many among them to whom the experience has come at some supreme moment of their lives, when it is comprehensible that there might have been a temporary exaltation of faculty which would be sufficient to account for it.

In the case of another subdivision of them the solitary case has been the seeing of an apparition, most commonly of some friend or relative at the point of death. Two possibilities are then offered for our choice, and in each of them the strong wish of the dying man is the impelling force. That force may have enabled him to materialize himself for a moment, in which case of course no clairvoyance was needed or more probably it may have acted mesmerically upon the percipient, and momentarily dulled his physical and stimulated his higher sensitiveness. In either case the vision is the product of the emergency, and is not repeated simply because the necessary conditions are not repeated.

There remains, however, an irresolvable residuum of cases in which a solitary instance occurs of the exercise of undoubted clairvoyance, while yet the occasion seems to us wholly trivial and unimportant. About these we can only frame hypotheses; the governing conditions are evidently not on the physical plane, and a separate investigation of each case would be necessary before we could speak with any certainty as to its causes. In some such it has appeared that an astral entity was endeavoring to make some communication, and was able to impress only some unimportant detail on its subject--all the useful or significant part of what it had to say failing to get through into the subject's consciousness.

In the investigation of the phenomena of clairvoyance all these varied types and many others will be encountered, and a certain number of cases of mere hallucination will be almost sure to appear also, and will have to be carefully weeded out from the list of examples. The student of such a subject needs an inexhaustible fund of patience and steady perseverance, but if he goes on long enough he will begin dimly to discern order behind the chaos, and will gradually get some idea of the great laws under which the whole evolution is working.

It will help him greatly in his efforts if he will adopt the order which we have just followed--that is, if he will first take the trouble to familiarize himself as thoroughly as may be with the actual facts concerning the planes with which ordinary clairvoyance deals. If he will learn what there really is to be seen with astral and etheric sight, and what their respective limitations are, he will then have, as it were, a standard by which to measure the cases which he observes. Since all instances of partial sight must of necessity fit into some niche in this whole, if he has the outline of the entire scheme in his head he will find it comparatively easy with a little practice to classify the instances with which he is called upon to deal.

We have said nothing as yet as to the still more wonderful possibilities of clairvoyance upon the mental plane, nor indeed is it necessary that much should be said, as it is exceedingly improbable that the investigator will ever meet with any examples of it except among pupils properly trained in some of the very highest schools of occultism. For them it opens up yet another new world, vaster far than all those beneath it--a world in which all that we can imagine of utmost glory and splendor is the commonplace of existence. Some account of its marvelous faculty, its ineffable bliss, its magnificent opportunities for learning and for work, is given in the sixth of our Theosophical manuals, and to that the student may be referred.

All that it has to give--all of it at least that he can assimilate--is within the reach of the trained pupil, but for the untrained clairvoyant to touch it is hardly more than a bare possibility. It has been done in mesmeric trance, but the occurrence is of exceeding rarity, for it needs almost superhuman qualifications in the way of lofty spiritual aspiration and absolute purity of thought and intention upon the part both of the subject and the operator.

To a type of clairvoyance such as this, and still more fully to that which belongs to the plane next above it, the name of spiritual sight may reasonably be applied; and since the celestial world to which it opens our eyes lies all round us here and now, it is fit that our passing reference to it should be made under the heading of simple clairvoyance, though it may be necessary to allude to it again when dealing with clairvoyance in space, to which we will now pass on.

CHAPTER 4

CLAIRVOYANCE IN SPACE: INTENTIONAL

We have defined this as the capacity to see events or scenes removed from the seer in space and too far distant for ordinary observation. The instances of this are so numerous and so various that we shall find it desirable to attempt a somewhat more detailed classification of them. It does not much matter what particular arrangement we adopt, so long as it is comprehensive enough to include all our cases; perhaps a convenient one will be to group them under the broad divisions of intentional and unintentional clairvoyance in space, with an intermediate class that might be described as semi-intentional--a curious title, but I will explain it later.

As before, I will begin by stating what is possible along this line for the fully-trained seer, and endeavoring to explain how his faculty works and under what limitations it acts. After that we shall find ourselves in a better position to try to understand the manifold examples of partial and untrained sight. Let us then in the first place discuss intentional clairvoyance.

It will be obvious from what has previously been said as to the power of astral vision that any one possessing it in its fullness will be able to see by its means practically anything in this world that he wishes to see. The most secret places are open to his gaze, and intervening obstacles have no existence for him, because of the change in his point of view; so that if we grant him the power of moving about in the astral body he can without difficulty go anywhere and see anything within the limits of the planet. Indeed this is to a large extent possible to him even without the necessity of moving the astral body at all, as we shall presently see.

Let us consider a little more closely the methods by which this super-physical sight may be used to observe events taking place at a distance. When, for example, a man here in England sees in minutest detail something which is happening at the same moment in India or America, how is it done?

A very ingenious hypothesis has been offered to account for the phenomenon. It has been suggested that every object is perpetually throwing off radiations in all directions, similar in some respects to, though infinitely finer than, rays of light, and that clairvoyance is nothing but the power to see by means of these finer radiations. Distance would in that case be no bar to the sight, all intervening objects would be penetrable by these rays, and they would be able to cross one another to infinity in all directions without entanglement, precisely as the vibrations of ordinary light do.

Now though this is not exactly the way in which clairvoyance works, the theory is nevertheless quite true in most of its premises. Every object undoubtedly is throwing off radiations in all directions, and it is precisely in this way, though on a higher plane, that the âkâshic records seem to be formed. Of them it will be necessary to say something under our next heading, so we will do no more than mention them for the moment. The phenomena of psychometry are also dependent upon these radiations, as will presently be explained.

There are, however, certain practical difficulties in the way of using these etheric vibrations (for that is, of course, what they are) as the medium by means of which one may see anything taking place at a distance. Intervening objects are not entirely transparent, and as the actors in the scene which the experimenter tried to observe would probably be at least equally transparent, it is obvious that serious confusion would be quite likely to result.

The additional dimension which would come into play if astral radiations were sensed instead of etheric would obviate some of the difficulties, but would on the other hand introduce some fresh complications of its own; so that for practical purposes, in endeavoring to understand clairvoyance, we may dismiss this hypothesis of radiations from our minds, and turn to the methods of seeing at a distance which are actually at the disposal of the student. It will be found that there are five, four of them being really varieties of clairvoyance, while the fifth does not properly come under that head at all, but belongs to the domain of magic. Let us take this last one first, and get it out of our way.

1. *By the assistance of a nature-spirit.*--This method does not necessarily involve the possession of any psychic faculty at all on the part of the experimenter; he need only know how to induce some denizen of the astral world to undertake the investigation for him. This may be done either by invocation or by evocation; that is to say, the operator may either persuade his astral coadjutor by prayers and offerings to give him the help he desires, or he may compel his aid by the determined exercise of a highly-developed will.

This method has been largely practiced in the East (where the entity employed is usually a nature-spirit) and in old Atlantis, where "the lords of the dark face" used a highly-specialized and peculiarly venomous variety of artificial elemental for this purpose. Information is sometimes obtained in the same sort of way at the spiritualistic *séance* of modern days, but in that case the messenger employed is more likely to be a recently-deceased human being functioning more or less freely on the astral plane-- though even here also it is sometimes an obliging nature-spirit, who is amusing himself by posing as somebody's departed relative. In any case, as I have said, this method is not clairvoyant at all, but magical; and it is mentioned here only in order that the reader may not become confused in the endeavor to classify cases of its use under some of the following headings.

2. *By means of an astral current.*--This is a phrase frequently and rather loosely employed in some of our Theosophical literature to cover a considerable variety of phenomena, and among others that which I wish to explain. What is really done by the student who adopts this method is not so much the setting in motion of a current in astral matter, as the erection of a kind of temporary telephone through it.

It is impossible here to give an exhaustive disquisition on astral physics, even had I the requisite knowledge to write it; all I need say is that it is possible to make in astral matter a definite connecting-line that shall act as a telegraph-wire to convey vibrations by means of which all that is going on at the other end of it may be seen. Such a line is established, be it understood, not by a direct projection through space of astral matter, but by such action upon a line (or rather many lines) of particles of that matter as will render them capable of forming a conductor for vibrations of the character required.

This preliminary action can be set up in two ways--either by the transmission of energy from particle to particle, until the line is formed, or by the use of a force from a higher plane which is capable of acting upon the whole line simultaneously. Of course this latter method implies far greater development, since it involves the knowledge of (and the power to use) forces of a considerably higher level; so that the man who could make his line in this way would not, for his own use, need a line at all, since he could see far more easily and completely by means of an altogether higher faculty.

Even the simpler and purely astral operation is a difficult one to describe, though quite an easy one to perform. It may be said to partake somewhat of the nature of the magnetization of a bar of steel; for it consists in what we might call the polarization, by an effort of the human will, of a number of parallel lines of astral atoms reaching from the operator to the scene which he wishes to observe. All the atoms thus affected are held for the time with their axes rigidly parallel to one another, so that they form a kind of temporary tube along which the clairvoyant may look. This method has the disadvantage that the telegraph line is liable to disarrangement or even destruction by any sufficiently strong astral current which happens to cross its path; but if the original effort of will were fairly definite, this would be a contingency of only infrequent occurrence.

The view of a distant scene obtained by means of this "astral current" is in many ways not unlike that seen through a telescope. Human figures usually appear very small, like those on a distant stage, but in spite of their diminutive size they are as clear as though they were close by. Sometimes it is possible by this means to hear what is said as well as to see what is done; but as in the majority of cases this does not happen, we must consider it rather as the manifestation of an additional power than as a necessary corollary of the faculty of sight.

It will be observed that in this case the seer does not usually leave his physical body at all; there is no sort of projection of his astral vehicle or of any part of himself towards that at which he is looking, but he simply manufactures for himself a temporary astral telescope. Consequently he has, to a certain extent, the use of his physical powers even while he is examining the distant scene; for example, his voice would usually still be under his control, so that he could describe what he saw even while he was in the act of making his observations. The consciousness of the man is, in fact, distinctly still at this end of the line.

This fact, however, has its limitations as well as its advantages, and these again largely resemble the limitations of the man using a telescope on the physical plane. The experimenter, for example, has no power to shift this point of view; his telescope, so to speak, has a particular field of view which cannot be enlarged or altered; he is looking at his scene from a certain direction, and he cannot suddenly turn it all round and see how it looks from the other side. If he has sufficient psychic energy to spare, he may drop altogether the telescope that he is using and manufacture an entirely new one for himself which will approach his objective somewhat differently; but this is not a course at all likely to be adopted in practice.

But, it may be said, the mere fact that he is using astral sight ought to enable him to see it from all sides at once. So it would if he were using that sight in the normal way upon an object which was fairly near him--within his astral reach, as it were; but at a distance of hundreds or thousands of miles the case is very different. Astral sight gives us the advantage of an additional dimension, but there is still such a thing as position in that dimension, and it is naturally a potent factor in limiting the use of the powers of its plane. Our ordinary three-dimensional sight enables us to see at once every point of the interior of a two-dimensional figure, such as a square, but in order to do that the square must be within a reasonable distance from our eyes; the mere additional dimension will avail a man in London but little in his endeavour to examine a square in Calcutta.

Astral sight, when it is cramped by being directed along what is practically a tube, is limited very much as physical sight would be under similar circumstances; though if possessed in perfection it will still continue to show, even at that distance, the auras, and therefore all the emotions and most of the thoughts of the people under observation.

There are many people for whom this type of clairvoyance is very much facilitated if they have at hand some physical object which can be used as a starting-point for their astral tube--a convenient focus for their will-power. A ball of crystal is the commonest and most effectual of such foci, since it has the additional advantage of possessing within itself qualities which stimulate psychic faculty; but other objects are also employed, to which we shall find it necessary to refer more particularly when we come to consider semi-intentional clairvoyance.

In connection with this astral-current form of clairvoyance, as with others, we find that there are some psychics who are unable to use it except when under the influence of mesmerism. The peculiarity in this case is that among such psychics there are two varieties--one in which by being thus set free the man is enabled to make a telescope for himself, and another in which the magnetizer himself makes the telescope and the subject is simply enabled to see through it. In this latter case obviously the subject has not enough will to form a tube for himself, and the operator, though possessed of the necessary will-power, is not clairvoyant, or he could see through his own tube without needing help.

Occasionally, though rarely, the tube which is formed possesses another of the attributes of a telescope--that of magnifying the objects at which it is directed until they seem of life-size. Of course the objects must always be magnified to some extent, or they would be absolutely invisible, but usually the extent is determined by the size of the astral tube, and the whole thing is simply a tiny moving picture. In the few cases where the figures are seen as of life-size by this method, it is probable that an altogether new power is beginning to dawn; but when this happens, careful observation is needed in order to distinguish them from examples of our next class.

3. *By the projection of a thought-form.*--The ability to use this method of clairvoyance implies a development somewhat more advanced than the last, since it necessitates a certain amount of control upon the mental plane. All students of Theosophy are aware that thought takes form, at any rate upon its own plane, and in the vast majority of cases upon the astral plane also; but it may not be quite so generally known that if a man thinks strongly of himself as present at any given place, the form assumed by that particular thought will be a likeness of the thinker himself, which will appear at the place in question.

Essentially this form must be composed of the matter of the mental plane, but in very many cases it would draw round itself matter of the astral plane also, and so would approach much nearer to visibility. There are, in fact, many instances in which it has been seen by the person thought of--most probably by means of the unconscious mesmeric influence emanating from the original thinker. None of the consciousness of the thinker would, however, be included within this thought-form. When once sent out from him, it would normally be a quite separate entity--not indeed absolutely unconnected with its maker, but practically so as far as the possibility of receiving any impression through it is concerned.

This third type of clairvoyance consists, then, in the power to retain so much connection with and so much hold over a newly-erected thought-form as will render it possible to receive impressions by means of it. Such impressions as were made upon the form would in this case be transmitted to the thinker--not along an astral telegraph line, as before, but by sympathetic vibration. In a perfect case of this kind of clairvoyance it is almost as though the seer projected a part of his consciousness into the thought-form, and used it as a kind of outpost, from which observation was possible. He sees almost as well as he would if he himself stood in the place of his thought-form.

The figures at which he is looking will appear to him as of life-size and close at hand, instead of tiny and at a distance, as in the previous case; and he will find it possible to shift his point of view if he wishes to do so. Clairaudience is perhaps less frequently associated with this type of clairvoyance than with the last, but its place is to some extent taken by a kind of mental perception of the thoughts and intentions of those who are seen.

Since the man's consciousness is still in the physical body, he will be able (even while exercising the faculty) to hear and to speak, in so far as he can do this without any distraction of his attention. The moment that the intentness of his thought fails the whole vision is gone, and he will have to construct a fresh thought-form before he can resume it. Instances in which this kind of sight is possessed with any degree of perfection by untrained people are naturally rarer than in the case of the previous type, because of the capacity for mental control required, and the generally finer nature of the forces employed.

4. *By travelling in the astral body.*--We enter here upon an entirely new variety of clairvoyance, in which the consciousness of the seer no longer remains in or closely connected with his physical body, but is definitely transferred to the scene which he is examining. Though it has no doubt greater dangers for the untrained seer than either of the methods previously described, it is yet quite the most satisfactory form of clairvoyance open to him, for the immensely superior variety which we shall consider under our fifth head is not available except for specially trained students.

In this case the man's body is either asleep or in trance, and its organs are consequently not available for use while the vision is going on, so that all description of what is seen, and all questioning as to further particulars, must be postponed until the wanderer returns to this plane. On the other hand the sight is much fuller and more perfect; the man hears as well as sees everything which passes before him, and can move about freely at will within the very wide limits of the astral plane. He can see and study at leisure all the other inhabitants of that plane, so that the great world of the nature-spirits (of which the traditional fairy-land is but a very small part) lies open before him, and even that of some of the lower devas.

He has also the immense advantage of being able to take part, as it were, in the scenes which come before his eyes--of conversing at will with these various astral entities, from whom so much information that is curious and interesting may be obtained. If in addition he can learn how to materialize himself (a matter of no

45

great difficulty for him when once the knack is acquired), he will be able to take part in physical events or conversations at a distance, and to show himself to an absent friend at will.

Again, he has the additional power of being able to hunt about for what he wants. By means of the varieties of clairvoyance previously described, for all practical purposes he could find a person or a place only when he was already acquainted with it, or when he was put *en rapport* with it by touching something physically connected with it, as in psychometry. It is true that by the third method a certain amount of motion is possible, but the process is a tedious one except for quite short distances.

By the use of the astral body, however, a man can move about quite freely and rapidly in any direction, and can (for example) find without difficulty any place pointed out upon a map, without either any previous knowledge of the spot or any object to establish a connection with it. He can also readily rise high into the air so as to gain a bird's-eye view of the country which he is examining, so as to observe its extent, the contour of its coast-line, or its general character. Indeed, in every way his power and freedom are far greater when he uses this method than they have been in any of the previous cases.

A good example of the full possession of this power is given, on the authority of the German writer Jung Stilling, by Mrs. Crowe in *The Night Side of Nature* (p. 127). The story is related of a seer who is stated to have resided in the neighborhood of Philadelphia, in America. His habits were retired, and he spoke little; he was grave, benevolent and pious, and nothing was known against his character except that he had the reputation of possessing some secrets that were considered not altogether *lawful.* Many extraordinary stories were told of him, and amongst the rest the following:--

"The wife of a ship captain (whose husband was on a voyage to Europe and Africa, and from whom she had been long without tidings), being overwhelmed with anxiety for his safety, was induced to address herself to this person. Having listened to her story he begged her to excuse him for a while, when he would bring her the intelligence she required. He then passed into an inner room and she sat herself down to wait; but his absence continuing longer than she expected, she became impatient, thinking he had forgotten her, and softly approaching the door she peeped through some aperture, and to her surprise beheld him lying on a sofa as motionless as if he were dead. She of course did not think it advisable to disturb him, but waited his return, when he told her that her husband had not been able to write to her for such and such reasons, but that he was then in a coffee-house in London and would very shortly be home again.

"Accordingly he arrived, and as the lady learnt from him that the causes of his unusual silence had been precisely those alleged by the man, she felt extremely desirous of ascertaining the truth of the rest of the information. In this she was gratified, for he no sooner set his eyes on the magician than he said that he had seen him before on a certain day in a coffee-house in London, and that he told him that his wife was extremely uneasy about him, and that he, the captain, had thereon mentioned how he had been prevented writing, adding that he was on the eve of embarking for America. He had then lost sight of the stranger amongst the throng, and knew nothing more about him."

We have of course no means now of knowing what evidence Jung Stilling had of the truth of this story, though he declares himself to have been quite satisfied with the authority on which he relates it; but so many similar things have happened that there is no reason to doubt its accuracy. The seer, however, must either have developed his faculty for himself or learnt it in some school other than that from which most of our Theosophical information is derived; for in our case there is a well-understood regulation expressly forbidding the pupils from giving any manifestation of such power which can be definitely proved at both ends in that way, and so constitute what is called "a phenomenon." That this regulation is emphatically a wise one is proved to all who know anything of the history of our Society by the disastrous results which followed from a very slight temporary relaxation of it.

I have given some quite modern cases almost exactly parallel to the above in my little book on *Invisible Helpers*. An instance of a lady well-known to myself, who frequently thus appears to friends at a distance, is given by Mr. Stead in *Real Ghost Stories* (p. 27); and

Mr. Andrew Lang gives, in his *Dreams and Ghosts* (p. 89), an account of how Mr. Cleave, then at Portsmouth, appeared intentionally on two occasions to a young lady in London, and alarmed her considerably. There is any amount of evidence to be had on the subject by any one who cares to study it seriously.

This paying of intentional astral visits seems very often to become possible when the principles are loosened at the approach of death for people who were unable to perform such a feat at any other time. There are even more examples of this class than of the other; I epitomize a good one given by Mr. Andrew Lang on p. 100 of the book last cited--one of which he himself says, "Not many stories have such good evidence in their favor."

"Mary, the wife of John Goffe of Rochester, being afflicted with a long illness, removed to her father's house at West Malling, about nine miles from her own.

"The day before her death she grew very impatiently desirous to see her two children, whom she had left at home to the care of a nurse. She was too ill to be moved, and between one and two o'clock in the morning she fell into a trance. One widow Turner, who watched with her that night, says that her eyes were open and fixed, and her jaw fallen. Mrs. Turner put her hand upon her mouth, but could perceive no breath. She thought her to be in a fit, and doubted whether she were dead or alive.

"The next morning the dying woman told her mother that she had been at home with her children, saying, I was with them last night when I was asleep.'

"The nurse at Rochester, widow Alexander by name, affirms that a little before two o'clock that morning she saw the likeness of the said Mary Goffe come out of the next chamber (where the elder child lay in a bed by itself), the door being left open, and stood by her bedside for about a quarter of an hour; the younger child was there lying by her. Her eyes moved and her mouth went, but she said nothing. The nurse, moreover, says that she was perfectly awake; it was then daylight, being one of the longest days in the year. She sat up in bed and looked steadfastly on the apparition. In that time she heard the bridge clock strike two, and a while after said: 'In the name of the Father, Son and Holy Ghost, what art thou?' Thereupon the apparition removed and went away; she slipped on her clothes and followed, but what became on't, she cannot tell."

The nurse apparently was more frightened by its disappearance than its presence, for after this she was afraid to stay in the house, and so spent the rest of the time until six o'clock in walking up and down outside. When the neighbors were awake she told her tale to them, and they of course said she had dreamt it all; she naturally enough warmly repudiated that idea, but could obtain no credence until the news of the other side of the story arrived from West Malling, when people had to admit that there might have been something in it.

A noteworthy circumstance in this story is that the mother found it necessary to pass from ordinary sleep into the profounder trance condition before she could consciously visit her children; it can, however, be paralleled here and there among the large number of similar accounts which may be found in the literature of the subject.

Two other stories of precisely the same type--in which a dying mother, earnestly desiring to see her children, falls into a deep sleep, visits them and returns to say that she has done so--are given by Dr. F. G. Lee. In one of them the mother, when dying in Egypt, appears to her children at Torquay, and is clearly seen in broad daylight by all five of the children and also by the nursemaid. (*Glimpses of the Supernatural*, vol. ii., p. 64.) In the other a Quaker lady dying at Cockermouth is clearly seen and recognized in daylight by her three children at Settle, the remainder of the story being practically identical with the one given above. (*Glimpses in the Twilight*, p. 94.) Though these cases appear to be less widely known than that of Mary Goffe, the evidence of their authenticity seems to be quite as good, as will be seen by the attestations obtained by the reverend author of the works from which they are quoted.

The man who fully possesses this fourth type of clairvoyance has many and great advantages at his disposal, even in addition to those already mentioned. Not only can he visit without trouble or expense all the beautiful and famous places of the earth, but if he happens to be a scholar, think what it must mean to him that he has access to all the libraries of the world! What must it be for the scientifically-minded man to see taking place before his eyes so many of the processes of the secret chemistry of nature, or for the philosopher to have revealed to him so much more than ever before of the working of the great mysteries of life and death? To him those who are gone from this plane are dead no longer, but living and within reach for a long time to come; for him many of the conceptions of religion are no longer matters of faith, but of knowledge. Above all, he can join the army of invisible helpers, and really be of use on a large scale. Undoubtedly clairvoyance, even when confined to the astral plane, is a great boon to the student.

Certainly it has its dangers also, especially for the untrained; danger from evil entities of various kinds, which may terrify or injure those who allow themselves to lose the courage to face them boldly; danger of deception of all sorts, of misconceiving and mis-interpreting what is seen; greatest of all, the danger of becoming conceited about the thing and of thinking it impossible to make a mistake. But a little common-sense and a little experience should easily guard a man against these.

5. *By travelling in the mental body.*--This is simply a higher and, as it were, glorified form of the last type. The vehicle employed is no longer the astral body, but the mind-body--a vehicle, therefore, belonging to the mental plane, and having within it all the potentialities of the wonderful sense of that plane, so transcendent in its action yet so impossible to describe. A man functioning in this leaves his astral body behind him along with the physical, and if he wishes to show himself upon the astral plane for any reason, he does not send for his own astral vehicle, but just by a single action of his will materializes one for his temporary need. Such an astral materialization is sometimes called the mâyâvirûpa, and to form it for the first time usually needs the assistance of a qualified Master.

The enormous advantages given by the possession of this power are the capacity of entering upon all the glory and the beauty of the higher land of bliss, and the possession, even when working on the astral plane, of the far more comprehensive mental sense which opens up to the student such marvelous vistas of knowledge, and practically renders error all but impossible. This higher flight, however, is possible for the trained man only, since only under definite training can a man at this stage of evolution learn to employ his mental body as a vehicle.

Before leaving the subject of full and intentional clairvoyance, it may be well to devote a few words to answering one or two questions as to its limitations, which constantly occur to students. Is it possible, we are often asked, for the seer to find any person with whom he wishes to communicate, anywhere in the world, whether he be living or dead?

To this reply must be a conditional affirmative. Yes, it is possible to find any person if the experimenter can, in some way or other, put himself *en rapport* with that person. It would be hopeless to plunge vaguely into space to find a total stranger among all the millions around us without any kind of clue; but, on the other hand, a very slight clue would usually be sufficient.

If the clairvoyant knows anything of the man whom he seeks, he will have no difficulty in finding him, for every man has what may be called a kind of musical chord of his own--a chord which is the expression of him as a whole, produced perhaps by a sort of average of the rates of vibration of all his different vehicles on their respective planes. If the operator knows how to discern that chord and to strike it, it will by sympathetic vibration attract the attention of the man instantly wherever he may be, and will evoke an immediate response from him.

Whether the man were living or recently dead would make no difference at all, and clairvoyance of the fifth class could at once find him even among the countless millions in the heaven-world, though in that case the man himself would be unconscious that he was under observation. Naturally a seer whose consciousness did not range higher than the astral plane--who employed therefore one of the earlier methods of seeing--would not be able to find a person upon the mental plane at all; yet even he would at least be able to tell that the man sought for was upon that plane, from the mere fact that the striking of the chord as far up as the astral level produced no response.

If the man sought be a stranger to the seeker, the latter will need something connected with him to act as a clue--a photograph, a letter written by him, an article which has belonged to him, and is impregnated with his personal magnetism; any of these would do in the hands of a practiced seer.

Again I say, it must not therefore be supposed that pupils who have been taught how to use this art are at liberty to set up a kind of intelligence office through which communication can be had with missing or dead relatives. A message given from this side to such an one might or might not be handed on, according to circumstances, but even if it were, no reply might be brought, lest the transaction should partake of the nature of a phenomenon--something which could be proved on the physical plane to have been an act of magic.

Another question often raised is as to whether, in the action of psychic vision, there is any limitation as to distance. The reply would seem to be that there should be no limit but that of the respective planes. It must be remembered that the astral and mental planes of our earth are as definitely its own as its atmosphere, though they extend considerably further from it even in our three-dimensional space than does the physical air. Consequently the passage to, or the detailed sight of, other planets would not be possible for any system of clairvoyance connected with these planes. It *is* quite possible and easy for the man who can raise his consciousness to the buddhic plane to pass to any other globe belonging to our chain of worlds, but that is outside our present subject.

Still a good deal of additional information about other planets can be obtained by the use of such clairvoyant faculties as we have been describing. It is possible to make sight enormously clearer by passing outside of the constant disturbances of the earth's atmosphere, and it is also not difficult to learn how to put on an exceedingly high magnifying power, so that even by ordinary clairvoyance a good deal of very interesting astronomical knowledge may be gained. But as far as this earth and its immediate surroundings are concerned, there is practically no limitation.

CHAPTER 5

CLAIRVOYANCE IN SPACE: SEMI-INTENTIONAL

Under this rather curious title I am grouping together the cases of all those people who definitely set themselves to see something, but have no idea what the something will be, and no control over the sight after the visions have begun--psychic Micawbers, who put themselves into a receptive condition, and then simply wait for something to turn up. Many trance-mediums would come under this heading; they either in some way hypnotize themselves or are hypnotized by some "spirit-guide," and then they describe the scenes or persons that happen to float before their vision. Sometimes, however, when in this condition they see what is taking place at a distance, and so they come to have a place among our "clairvoyants in space."

But the largest and most widely-spread band of these semi-intentional clairvoyants are the various kinds of crystal-gazers-- those who, as Mr. Andrew Lang puts it, "stare into a crystal ball, a cup, a mirror, a blob of ink (Egypt and India), a drop of blood (among the Maories of New Zealand), a bowl of water (Red Indian), a pond (Roman and African), water in a glass bowl (in Fez), or almost any polished surface" (*Dreams and Ghosts*, p. 57).

Two pages later Mr. Lang gives us a very good example of the kind of vision most frequently seen in this way. "I had given a glass ball," he says, "to a young lady, Miss Baillie, who had scarcely any success with it. She lent it to Miss Leslie, who saw a large square, old-fashioned red sofa covered with muslin, which she found in the next country-house she visited. Miss Baillie's brother, a young athlete, laughed at these experiments, took the ball into the study, and came back looking 'gey gash.' He admitted that he had seen a vision--somebody he knew under a lamp. He would discover during the week whether he saw right or not. This was at 5.30 on a Sunday afternoon.

"On Tuesday, Mr. Baillie was at a dance in a town some forty miles from his home, and met a Miss Preston. 'On Sunday,' he said, 'about half-past five you were sitting under a standard lamp in a dress I never saw you wear, a blue blouse with lace over the shoulders, pouring out tea for a man in blue serge, whose back was towards me, so that I only saw the tip of his moustache.'

"'Why, the blinds must have been up,' said Miss Preston.

"'I was at Dulby,' said Mr. Baillie, and he undeniably was."

This is quite a typical case of crystal-gazing--the picture correct in every detail, you see, and yet absolutely unimportant and bearing no apparent signification of any sort to either party, except that it served to prove to Mr. Baillie that there was something in crystal-gazing. Perhaps more frequently the visions tend to be of a romantic character--men in foreign dress, or beautiful though generally unknown landscapes.

Now what is the rationale of this kind of clairvoyance? As I have indicated above, it belongs usually to the "astral-current" type, and the crystal or other object simply acts as a focus for the will-power of the seer, and a convenient starting-point for his astral tube. There are some who can influence what they will see by their will, that is to say they have the power of pointing their telescope as they wish; but the great majority just form a fortuitous tube and see whatever happens to present itself at the end of it.

Sometimes it may be a scene comparatively near at hand, as in the case just quoted; at other times it will be a far-away Oriental landscape; at others yet it may be a reflection of some fragment of an âkâshic record, and then the picture will contain figures in some antique dress, and the phenomenon belongs to our third large division of "clairvoyance in time." It is said that visions of the future are sometimes seen in crystals also--a further development to which we must refer later.

I have seen a clairvoyant use instead of the ordinary shining surface a dead black one, produced by a handful of powdered charcoal in a saucer. Indeed it does not seem to matter much what is used as a focus, except that pure crystal has an undoubted advantage over other substances in that its peculiar arrangement of elemental essence renders it specially stimulating to the psychic faculties.

It seems probable, however, that in cases where a tiny brilliant object is employed--such as a point of light, or the drop of blood used by the Maories--the instance is in reality merely one of self-hypnotization. Among non-European nations the experiment is very frequently preceded or accompanied by magical ceremonies and invocations, so that it is quite likely that such sight as is gained may sometimes be really that of some foreign entity, and so the phenomenon may in fact be merely a case of temporary possession, and not of clairvoyance at all.

CHAPTER 6

CLAIRVOYANCE IN SPACE: UNINTENTIONAL

Under this heading we may group together all those cases in which visions of some event which is taking place at a distance are seen quite unexpectedly and without any kind of preparation. There are people who are subject to such visions, while there are many others to whom such a thing will happen only once in a life-time. The visions are of all kinds and of all degrees of completeness, and apparently may be produced by various causes. Sometimes the reason of the vision is obvious, and the subject matter of the gravest importance; at other times no reason at all is discoverable, and the events shown seem of the most trivial nature.

Sometimes these glimpses of the super-physical faculty come as waking visions, and sometimes they manifest during sleep as vivid or oft-repeated dreams. In this latter case the sight employed is perhaps usually of the kind assigned to our fourth subdivision of clairvoyance in space, for the sleeping man often travels in his astral body to some spot with which his affections or interests are closely connected, and simply watches what takes place there; in the former it seems probable that the second type of clairvoyance, by means of the astral current, is called into requisition. But in this case the current or tube is formed quite unconsciously, and is often the automatic result of a strong thought or emotion projected from one end or the other--either from the seer or the person who is seen.

The simplest plan will be to give a few instances of the different kinds, and to intersperse among them such further explanations as may seem necessary. Mr. Stead has collected a large and varied assortment of recent and well-authenticated cases in his *Real*

Ghost Stories, and I will select some of my examples from them, occasionally condensing slightly to save space.

There are cases in which it is at once obvious to any Theosophical student that the exceptional instance of clairvoyance was specially brought about by one of the band whom we have called "Invisible Helpers" in order that aid might be rendered to some one in sore need. To this class, undoubtedly, belongs the story told by Captain Yonnt, of the Napa Valley in California, to Dr. Bushnell, who repeats it in his *Nature and the Supernatural* (p. 14).

"About six or seven years previous, in a mid-winter's night, he had a dream in which he saw what appeared to be a company of emigrants arrested by the snows of the mountains, and perishing rapidly by cold and hunger. He noted the very cast of the scenery, marked by a huge, perpendicular front of white rock cliff; he saw the men cutting off what appeared to be tree-tops rising out of deep gulfs of snow; he distinguished the very features of the persons and the look of their particular distress.

"He awoke profoundly impressed by the distinctness and apparent reality of the dream. He at length fell asleep, and dreamed exactly the same dream over again. In the morning he could not expel it from his mind. Falling in shortly after with an old hunter comrade, he told his story, and was only the more deeply impressed by his recognizing without hesitation the scenery of the dream. This comrade came over the Sierra by the Carson Valley Pass, and declared that a spot in the Pass exactly answered his description.

"By this the unsophistical patriarch was decided. He immediately collected a company of men, with mules and blankets and all necessary provisions. The neighbors were laughing meantime at his credulity. 'No matter,' he said, 'I am able to do this, and I will, for I verily believe that the fact is according to my dream.' The men were sent into the mountains one hundred and fifty miles distant direct to the Carson Valley Pass. And there they found the company exactly in the condition of the dream, and brought in the remnant alive."

Since it is not stated that Captain Yonnt was in the habit of seeing visions, it seems clear that some helper, observing the forlorn condition of the emigrant party, took the nearest impressionable and otherwise suitable person (who happened to be the Captain) to the spot in the astral body, and aroused him sufficiently to fix the scene firmly in his memory. The helper may possibly have arranged an "astral current" for the Captain instead, but the former suggestion is more probable. At any rate the motive, and broadly the method, of the work are obvious enough in this case.

Sometimes the "astral current" may be set going by a strong emotional thought at the other end of the line, and this may happen even though the thinker has no such intention in his mind. In the rather striking story which I am about to quote, it is evident that the link was formed by the doctor's frequent thought about Mrs. Broughton, yet he had clearly no especial wish that she should see what he was doing at the time. That it was this kind of clairvoyance that was employed is shown by the fixity of her point of view--which, be it observed, is not the doctor's point of view sympathetically transferred (as it might have been) since she sees his back without recognizing him. The story is to be found in the *Proceedings of the Psychical Research Society* (vol. ii., p. 160).

"Mrs. Broughton awoke one night in 1844, and roused her husband, telling him that something dreadful had happened in France. He begged her to go to sleep again, and not trouble him. She assured him that she was not asleep when she saw what she insisted on telling him--what she saw in fact.

"First a carriage accident--which she did not actually see, but what she saw was the result--a broken carriage, a crowd collected, a figure gently raised and carried into the nearest house, then a figure lying on a bed which she then recognized as the Duke of Orleans. Gradually friends collecting round the bed--among them several members of the French royal family--the queen, then the king, all silently, tearfully, watching the evidently dying duke. One man (she could see his back, but did not know who he was) was a doctor. He stood bending over the duke, feeling his pulse, with his watch in the other hand. And then all passed away, and she saw no more.

"As soon as it was daylight she wrote down in her journal all that she had seen. It was before the days of electric telegraph, and two or more days passed before the *Times* announced 'The Death of the Duke of Orleans.' Visiting Paris a short time afterwards she saw and recognized the place of the accident and received the explanation of her impression. The doctor who attended the dying duke was an old friend of hers, and as he watched by the bed his mind had been constantly occupied with her and her family."

A commoner instance is that in which strong affection sets up the necessary current; probably a fairly steady stream of mutual thought is constantly flowing between the two parties in the case, and some sudden need or dire extremity on the part of one of them endues this stream temporarily with the polarizing power which is needful to create the astral telescope. An illustrative example is quoted from the same *Proceedings* (vol. i., p. 30).

"On September 9th, 1848, at the siege of Mooltan, Major-General R----, C.B., then adjutant of his regiment, was most severely and dangerously wounded; and, supposing himself to be dying, asked one of the officers with him to take the ring off his finger and send it to his wife, who at the time was fully one hundred and fifty miles distant at Ferozepore.

"'On the night of September 9th, 1848,' writes his wife, 'I was lying on my bed, between sleeping and waking, when I distinctly saw my husband being carried off the field seriously wounded, and heard his voice saying, "Take this ring off my finger and send it to my wife." All the next day I could not get the sight or the voice out of my mind.

"'In due time I heard of General R..... having been severely wounded in the assault of Mooltan. He survived, however, and is still living. It was not for some time after the siege that I heard from General L----, the officer who helped to carry my husband off the field, that the request as to the ring was actually made by him, just as I heard it at Ferozepore at that very time."

Then there is the very large class of casual clairvoyant visions which have no traceable cause--which are apparently quite meaningless, and have no recognizable relation to any events known to the seer. To this class belong many of the landscapes seen by some people just before they fall asleep. I quote a capital and very realistic account of an experience of this sort from Mr. W. T. Stead's *Real Ghost Stories* (p. 65).

"I got into bed but was not able to go to sleep. I shut my eyes and waited for sleep to come; instead of sleep, however, there came to me a succession of curiously vivid clairvoyant pictures. There was no light in the room, and it was perfectly dark; I had my eyes shut also. But notwithstanding the darkness I suddenly was conscious of looking at a scene of singular beauty. It was as if I saw a living miniature about the size of a magic-lantern slide. At this moment I can recall the scene as if I saw it again. It was a seaside piece. The moon was shining upon the water, which rippled slowly on to the beach. Right before me a long mole ran into the water.

"On either side of the mole irregular rocks stood up above the sea-level. On the shore stood several houses, square and rude, which resembled nothing that I had ever seen in house architecture. No one was stirring, but the moon was there and the sea and the gleam of the moonlight on the rippling waters, just as if I had been looking on the actual scene.

"It was so beautiful that I remember thinking that if it continued I should be so interested in looking at it that I should never go to sleep. I was wide awake, and at the same time that I saw the scene I distinctly heard the dripping of the rain outside the window. Then suddenly, without any apparent object or reason, the scene changed.

"The moonlit sea vanished, and in its place I was looking right into the interior of a reading-room. It seemed as if it had been used as a schoolroom in the daytime, and was employed as a reading-room in the evening. I remember seeing one reader who had a curious resemblance to Tim Harrington, although it was not he, hold up a magazine or book in his hand and laugh. It was not a picture--it was there.

"The scene was just as if you were looking through an opera-glass; you saw the play of the muscles, the gleaming of the eye, every movement of the unknown persons in the unnamed place into which you were gazing. I saw all that without opening my eyes, nor did my eyes have anything to do with it. You see such things as these as it were with another sense which is more inside your head than in your eyes.

"This was a very poor and paltry experience, but it enabled me to understand better how it is that clairvoyants see than any amount of disquisition.

"The pictures were *apropos* of nothing; they had been suggested by nothing I had been reading or talking of; they simply came as if I had been able to look through a glass at what was occurring somewhere else in the world. I had my peep, and then it passed, nor have I had a recurrence of a similar experience."

Mr. Stead regards that as a "poor and paltry experience," and it may perhaps be considered so when compared with the greater possibilities, yet I know many students who would be very thankful to have even so much of direct personal experience to tell. Small though it may be in itself, it at once gives the seer a clue to the whole thing, and clairvoyance would be a living actuality to a man who had seen even that much in a way that it could never have been without that little touch with the unseen world.

These pictures were much too clear to have been mere reflections of the thought of others, and besides, the description unmistakably shows that they were views seen through an astral telescope; so either Mr. Stead must quite unconsciously have set a current going for himself, or (which is much more probable) some kindly astral entity set it in motion for him, and gave him, to while away a tedious delay, any pictures that happened to come handy at the end of the tube.

CHAPTER 6

CLAIRVOYANCE IN TIME: THE PAST

Clairvoyance in time--that is to say, the power of reading the past and the future--is, like all the other varieties, possessed by different people in very varying degrees, ranging from the man who has both faculties fully at his command, down to one who only occasionally gets involuntary and very imperfect glimpses or reflections of these scenes of other days. A person of the latter type might have, let us say, a vision of some event in the past; but it would be liable to the most serious distortion, and even if it happened to be fairly accurate it would almost certainly be a mere isolated picture, and he would probably be quite unable to relate it to what had occurred before or after it, or to account for anything unusual which might appear in it. The trained man, on the other hand, could follow the drama connected with his picture backwards or forwards to any extent that might seem desirable, and trace out with equal ease the causes which had led up to it or the results which it in turn would produce.

We shall probably find it easier to grasp this somewhat difficult section of our subject if we consider it in the subdivisions which naturally suggest themselves, and deal first with the vision which looks backwards into the past, leaving for later examination that which pierces the veil of the future. In each case it will be well for us to try to understand what we can of the *modus operandi,* even though our success can at best be only a very modified one, owing first to the imperfect information on some parts of the subject at present possessed by our investigators, and secondly to the ever-recurring failure of physical words to express a hundredth part even of the little we do know about higher planes and faculties.

In the case then of a detailed vision of the remote past, how is it obtained, and to what plane of nature does it really belong? The answer to both these questions is contained in the reply that it is read from the âkâshic records; but that statement in return will require a certain amount of explanation for many readers. The word is in truth somewhat of a misnomer, for though the records are undoubtedly read from the âkâsha, or matter of the mental plane, yet it is not to it that they really belong. Still worse is the alternative title, "records of the astral light," which has sometimes been employed, for these records lie far beyond the astral plane, and all that can be obtained on it are only broken glimpses of a kind of double reflection of them, as will presently be explained.

Like so many others of our Theosophical terms, the word âkâsha has been very loosely used. In some of our earlier books it was considered as synonymous with astral light, and in others it was employed to signify any kind of invisible matter, from mûlaprakriti down to the physical ether. In later books its use has been restricted to the matter of the mental plane, and it is in that sense that the records may be spoken of as âkâshic, for although they are not originally made on that plane any more than on the astral, yet it is there that we first come definitely into contact with them and find it possible to do reliable work with them.

This subject of the records is by no means an easy one to deal with, for it is one of that numerous class which requires for its perfect comprehension faculties of a far higher order than any which humanity has yet evolved. The real solution of its problems lies on planes far beyond any that we can possibly know at present, and any view that we take of it must necessarily be of the most imperfect character, since we cannot but look at it from below instead of from above. The idea which we form of it must therefore be only partial, yet it need not mislead us unless we allow ourselves to think of the tiny fragment which is all that we can see as though it were the perfect whole. If we are careful that such conceptions as we may form shall be accurate as far as they go, we shall have nothing to unlearn, though much to add, when in the course of our further progress we gradually acquire the higher wisdom. Be it understood then at the commencement that a thorough grasp of our subject is an impossibility at the present stage of our evolution, and that many points will arise as to which no exact explanation is yet obtainable, though it may often be possible to suggest analogies and to indicate the lines along which an explanation must lie.

Let us then try to carry back our thoughts to the beginning of this solar system to which we belong. We are all familiar with the ordinary astronomical theory of its origin--that which is commonly called the nebular hypothesis--according to which it first came into existence as a gigantic glowing nebula, of a diameter far exceeding that of the orbit of even the outermost of the planets, and then, as in the course of countless ages that enormous sphere gradually cooled and contracted, the system as we know it was formed.

Occult science accepts that theory, in its broad outline, as correctly representing the purely physical side of the evolution of our system, but it would add that if we confine our attention to this physical side only we shall have a very incomplete and incoherent idea of what really happened. It would postulate, to begin with, that the exalted Being who undertakes the formation of a system (whom we sometimes call the Logos of the system) first of all forms in His mind a complete conception of the whole of it with all its successive chains of worlds. By the very act of forming that conception He calls the whole into simultaneous objective existence on the plane of His thought--a plane of course far above all those of which we know anything--from which the various globes descend when required into whatever state of further objectivity may be respectively destined for them. Unless we constantly bear in mind this fact of the real existence of the whole system from the very beginning on a higher plane, we shall be perpetually misunderstanding the physical evolution which we see taking place down here.

But occultism has more than this to teach us on the subject. It tells us not only that all this wonderful system to which we belong is called into existence by the Logos, both on lower and on higher planes, but also that its relation to Him is closer even than that, for it is absolutely a part of Him--a partial expression of Him upon the physical plane--and that the movement and energy of the whole system is *His* energy, and is all carried on within the limits of His aura. Stupendous as this conception is, it will yet not be wholly unthinkable to those of us who have made any study of the subject of the aura.

We are familiar with the idea that as a person progresses on the upward path his causal body, which is the determining limit of his aura, distinctly increases in size as well as in luminosity and purity of color. Many of us know from experience that the aura of a pupil who has already made considerable advance on the Path is very much larger than that of one who is but just setting his foot upon its first step, while in the case of an Adept the proportional increase is far greater still. We read in quite exoteric Oriental scriptures of the immense extension of the aura of the Buddha; I think that three miles is mentioned on one occasion as its limit, but whatever the exact measurement may be, it is obvious that we have here another record of this fact of the extremely rapid growth of the causal body as man passes on his upward way. There can be little doubt that the rate of this growth would itself increase in geometrical progression, so that it need not surprise us to hear of an Adept on a still higher level whose aura is capable of including the entire world at once; and from this we may gradually lead our minds up to the conception that there is a Being so exalted as to comprehend within Himself the whole of our solar system. And we should remember that, enormous as this seems to us, it is but as the tiniest drop in the vast ocean of space.

So of the Logos (who has in Him all the capacities and qualities with which we can possibly endow the highest God we can imagine) it is literally true, as was said of old, that "of Him and through Him, and to Him are all things," and "in Him we live and move and have our being."

Now if this be so, it is clear that whatever happens within our system happens absolutely within the consciousness of its Logos, and so we at once see that the true record must be His memory; and furthermore, it is obvious that on whatever plane that wondrous memory exists, it cannot but be far above anything that we know, and consequently whatever records we may find ourselves able to read must be only a reflection of that great dominant fact, mirrored in the denser media of the lower planes.

On the astral plane it is at once evident that this is so--that what we are dealing with is only a reflection of a reflection, and an exceedingly imperfect one, for such records as can be reached there are fragmentary in the extreme, and often seriously distorted. We know how universally water is used as a symbol of the astral light, and in this particular case it is a remarkably apt one. From the surface of still water we may get a clear reflection of the surrounding objects, just as from a mirror; but at the best it is only a reflection--a representation in two dimensions of three-dimensional objects, and therefore differing in all its qualities, except color, from that which it represents; and in addition to this, it is always reversed.

But let the surface of the water be ruffled by the wind and what do we find then? A reflection still, certainly, but so broken up and distorted as to be quite useless or even misleading as a guide to the shape and real appearance of the objects reflected. Here and there for a moment we might happen to get a clear reflection of some minute part of the scene--of a single leaf from a tree, for example; but it would need long labour and considerable knowledge of natural laws to build up anything like a true conception of the object reflected by putting together even a large number of such isolated fragments of an image of it.

Now in the astral plane we can never have anything approaching to what we have imaged as a still surface, but on the contrary we have always to deal with one in rapid and bewildering motion; judge, therefore, how little we can depend upon getting a clear and definite reflection. Thus a clairvoyant who possesses only the faculty of astral sight can never rely upon any picture of the past that comes before him as being accurate and perfect; here and there some part of it *may* be so, but he has no means of knowing which it is. If he is under the care of a competent teacher he may, by long and careful training, be shown how to distinguish between reliable and unreliable impressions, and to construct from the broken reflections some kind of image of the object reflected; but usually long before he has mastered those difficulties he will have developed the mental sight, which renders such labour unnecessary.

On the next plane, which we call the mental, conditions are very different. There the record is full and accurate, and it would be impossible to make any mistake in the reading. That is to say, if three clairvoyants possessing the powers of the mental plane agreed to examine a certain record there, what would be presented to their vision would be absolutely the same reflection in each case, and each would acquire a correct impression from it in reading it. It does not however follow that when they all compared notes later on the physical plane their reports would agree exactly. It is well known that if three people who witness an occurrence down here in the physical world set to work to describe it afterwards, their accounts will differ considerably, for each will have noticed especially those items which most appeal to him, and will insensibly have made them the prominent features of the event, sometimes ignoring other points which were in reality much more important.

Now in the case of an observation on the mental plane this personal equation would not appreciably affect the impressions received, for since each would thoroughly grasp the entire subject it would be impossible for him to see its parts out of due proportion; but, except in the case of carefully trained and experienced persons, this factor does come into play in transferring the impressions to the lower planes. It is in the nature of things impossible that any account given down here of a vision or experience on the mental plane can be complete, since nine-tenths of what is seen and felt there could not be expressed by physical words at all; and, since all expression must therefore be partial, there is obviously some possibility of selection as to the part expressed. It is for this reason that in all our Theosophical investigations of recent years so much stress has been laid upon the constant checking and verifying of clairvoyant testimony, nothing which rests upon the vision of one person only having been allowed to appear in our later books.

But even when the possibility of error from this factor of personal equation has been reduced to a minimum by a careful system of counter-checking, there still remains the very serious difficulty which is inherent in the operation of bringing down impressions from a higher plane to a lower one. This is something analogous to the difficulty experienced by a painter in his endeavour to reproduce a three-dimensional landscape on a flat surface--that is, practically in two dimensions. Just as the artist needs long and careful training of eye and hand before he can produce a satisfactory representation of nature, so does the clairvoyant need long and careful training before he can describe accurately on a lower plane what he sees on a higher one; and the probability of getting an exact description from an untrained person is about equal to that of getting a perfectly-finished landscape from one who has never learnt how to draw.

It must be remembered, too, that the most perfect picture is in reality infinitely far from being a reproduction of the scene which it represents, for hardly a single line or angle in it can ever be the same as those in the object copied. It is simply a very ingenious attempt to make upon one only of our five senses, by means of lines and colours on a flat surface, an impression similar to that which would have been made if we had actually had before us the scene depicted. Except by a suggestion dependent entirely on our own previous experience, it can convey to us nothing of the roar of the sea, of the scent of the flowers, of the taste of the fruit, or of the softness or hardness of the surface drawn.

Of exactly similar nature, though far greater in degree, are the difficulties experienced by a clairvoyant in his attempt to describe upon the physical plane what he has seen upon the astral; and they are furthermore greatly enhanced by the fact that, instead of having merely to recall to the minds of his hearers conceptions with which they are already familiar, as the artist does when he paints men or animals, fields or trees, he has to endeavor by the very imperfect means at his disposal to suggest to them conceptions which in most cases are absolutely new to them.

Small wonder then that, however vivid and striking his descriptions may seem to his audience, he himself should constantly be impressed with their total inadequacy, and should feel that his best efforts have entirely failed to convey any idea of what he really sees. And we must remember that in the case of the report given down here of a record read on the mental plane, this difficult operation of transference from the higher to the lower has taken place not once but twice, since the memory has been brought through the intervening astral plane. Even in a case where the investigator has the advantage of having developed his mental faculties so that he has the use of them while awake in the physical body, he is still hampered by the absolute incapacity of physical language to express what he sees.

Try for a moment to realize fully what is called the fourth dimension, of which we said something in an earlier chapter. It is easy enough to think of our own three dimensions--to image in our minds the length, breadth and height of any object; and we see that each of these three dimensions is expressed by a line at right angles to both of the others. The idea of the fourth dimension is that it might be possible to draw a fourth line which shall be at right angles to all three of those already existing.

Now the ordinary mind cannot grasp this idea in the least, though some few who have made a special study of the subject have gradually come to be able to realize one or two very simple four-dimensional figures. Still, no words that they can use on this plane can bring any image of these figures before the minds of others, and if any reader who has not specially trained himself along that line will make the effort to visualize such a shape he will find it quite impossible. Now to express such a form clearly in physical words would be, in effect, to describe accurately a single object on the astral plane; but in examining the records on the mental plane we should have to face the additional difficulties of a fifth dimension! So that the impossibility of fully explaining these records will be obvious to even the most superficial observation.

We have spoken of the records as the memory of the Logos, yet they are very much more than a memory in an ordinary sense of the word. Hopeless as it may be to imagine how these images appear from His point of view, we yet know that as we rise higher and higher we must be drawing nearer to the true memory--must be seeing more nearly as He sees; so that great interest attaches to the experience of the clairvoyant with reference to these records when he stands upon the buddhic plane--the highest which his consciousness can reach even when away from the physical body until he attains the level of the Arhats.

Here time and space no longer limit him; he no longer needs, as on the mental plane, to pass a series of events in review, for past, present and future are all alike simultaneously present to him, meaningless as that sounds down here. Indeed, infinitely below the

70

consciousness of the Logos as even that exalted plane is, it is yet abundantly clear from what we see there that to Him the record must be far more than what we call a memory, for all that has happened in the past and all that will happen in the future is *happening now* before His eyes just as are the events of what we call the present time. Utterly incredible, wildly incomprehensible, of course, to our limited understanding; yet absolutely true for all that.

Naturally we could not expect to understand at our present stage of knowledge how so marvelous a result is produced, and to attempt an explanation would only be to involve ourselves in a mist of words from which we should gain no real information. Yet a line of thought recurs to my mind which perhaps suggests the direction in which it is possible that that explanation may lie: and whatever helps us to realize that so astounding a statement may after all not be wholly impossible will be of assistance in broadening our minds.

Some thirty years ago I remember reading a very curious little book, called, I think, *The Stars and the Earth*, the object of which was to endeavor to show how it was scientifically possible that to the mind of God the past and the present might be absolutely simultaneous. Its arguments struck me at the time as decidedly ingenious, and I will proceed to summarize them, as I think they will be found somewhat suggestive in connection with the subject which we have been considering.

When we see anything, whether it be the book which we hold in our hands or a star millions of miles away, we do so by means of a vibration in the ether, commonly called a ray of light, which passes from the object seen to our eyes. Now the speed with which this vibration passes is so great--about 186,000 miles in a second--that when we are considering any object in our own world we may regard it as practically instantaneous. When, however, we come to deal with interplanetary distances we have to take the speed of light into consideration, for an appreciable period is occupied in traversing these vast spaces. For example it takes eight minutes and a quarter for light to travel to us from the sun, so that when we look at the solar orb we see it by means of a ray of light which left it more than eight minutes ago.

From this follows a very curious result. The ray of light by which we see the sun can obviously report to us only the state of affairs which existed in that luminary when it started on its journey, and would not be in the least affected by anything that happened there after it left; so that we really see the sun not as he *is*, but as he was eight minutes ago. That is to say that if anything important took place in the sun--the formation of a new sun-spot, for instance-- an astronomer who was watching the orb through his telescope at the time would be quite unaware of the incident while it was happening, since the ray of light bearing the news would not reach him until more than eight minutes later.

The difference is more striking when we consider the fixed stars, because in their case the distances are so enormously greater. The pole star, for example, is so far off that light, travelling at the inconceivable speed above mentioned, takes a little more than fifty years to reach our eyes; and from that follows the strange but inevitable inference that we see the pole star not as and where it is at this moment, but as and where it was fifty years ago. Nay, if to-morrow some cosmic catastrophe were to shatter the pole star into fragments, we should still see it peacefully shining in the sky all the rest of our lives; our children would grow up to middle age and gather their children about them in turn before the news of that tremendous accident reached any terrestrial eye. In the same way there are other stars so far distant that light takes thousands of years to travel from them to us, and with reference to their condition our information is therefore thousands of years behind time.

Now carry the argument a step farther. Suppose that we were able to place a man at the distance of 186,000 miles from the earth, and yet to endow him with the wonderful faculty of being able from that distance to see what was happening here as clearly as though he were still close beside us. It is evident that a man so placed would see everything a second after the time when it really happened, and so at the present moment he would be seeing what happened a second ago. Double the distance, and he would be two seconds behind time, and so on; remove him to the distance of the sun (still allowing him to preserve the same mysterious power of sight) and he would look down and watch you doing not what you *are* doing now, but what you *were* doing eight minutes and a quarter ago. Carry him away to the pole star, and he would see passing before his eyes the events of fifty years ago; he would be watching the childish gambols of those who at the very same moment were really middle-aged men. Marvellous as this may sound, it is literally and scientifically true, and cannot be denied.

The little book went on to argue logically enough that God, being almighty, must possess the wonderful power of sight which we have been postulating for our observer; and further, that being omnipresent, He must be at each of the stations which we mentioned, and also at every intermediate point, not successively but simultaneously. Granting these premises, the inevitable deduction follows that everything which has ever happened from the very beginning of the world *must* be at this very moment taking place before the eye of God--not a mere memory of it, but the actual occurrence itself being now under His observation.

All this is materialistic enough, and on the plane of purely physical science, and we may therefore be assured that it is *not* the way in which the memory of the Logos acts; yet it is neatly worked out and absolutely incontrovertible, and as I have said before, it is not without its use, since it gives us a glimpse of some possibilities which otherwise might not occur to us.

But, it may be asked, how is it possible, amid the bewildering confusion of these records of the past, to find any particular picture when it is wanted? As a matter of fact, the untrained clairvoyant usually cannot do so without some special link to put him *en rapport* with the subject required. Psychometry is an instance in point, and it is quite probable that our ordinary memory is really only another presentment of the same idea. It seems as though there were a sort of magnetic attachment or affinity between any particle of matter and the record which contains its history--an affinity which enables it to act as a kind of conductor between that record and the faculties of anyone who can read it.

For example, I once brought from Stonehenge a tiny fragment of stone, not larger than a pin's head, and on putting this into an envelope and handing it to a psychometer who had no idea what it was, she at once began to describe that wonderful ruin and the desolate country surrounding it, and then went on to picture vividly what were evidently scenes from its early history, showing that that infinitesimal fragment had been sufficient to put her into communication with the records connected with the spot from which it came. The scenes through which we pass in the course of our life seem to act in the same manner upon the cells of our brain as did the history of Stonehenge upon that particle of stone: they establish a connection with those cells by means of which our mind is put *en rapport* with that particular portion of the records, and so we "remember" what we have seen.

Even a trained clairvoyant needs some link to enable him to find the record of an event of which he has no previous knowledge. If, for example, he wished to observe the landing of Julius Caesar on the shores of England, there are several ways in which he might approach the subject. If he happened to have visited the scene of the occurrence, the simplest way would probably be to call up the image of that spot, and then run back through its records until he reached the period desired. If he had not seen the place, he might run back in time to the date of the event, and then search the Channel for a fleet of Roman galleys; or he might examine the records of Roman life at about that period, where he would have no difficulty in identifying so prominent a figure as Cæsar, or in tracing him when found through all his Gallic wars until he set his foot upon British land.

People often enquire as to the aspect of these records--whether they appear near or far away from the eye, whether the figures in them are large or small, whether the pictures follow one another as in a panorama or melt into one another like dissolving views, and so on. One can only reply that their appearance varies to a certain extent according to the conditions under which they are seen. Upon the astral plane the reflection is most often a simple picture, though occasionally the figures seen would be endowed with motion; in this latter case, instead of a mere snapshot a rather longer and more perfect reflection has taken place.

On the mental plane they have two widely different aspects. When the visitor to that plane is not thinking specially of them in any way, the records simply form a background to whatever is going on, just as the reflections in a pier-glass at the end of a room might form a background to the life of the people in it. It must always be borne in mind that under these conditions they are really merely reflections from the ceaseless activity of a great Consciousness upon a far higher plane, and have very much the appearance of an endless succession of the recently invented *cinematographe*, or living photographs. They do not melt into one another like dissolving views, nor do a series of ordinary pictures follow one another; but the action of the reflected figures constantly goes on, as though one were watching the actors on a distant stage.

But if the trained investigator turns his attention specially to any one scene, or wishes to call it up before him, an extraordinary change at once takes place, for this is the plane of thought, and to think of anything is to bring it instantaneously before you. For example, if a man wills to see the record of that event to which we before referred--the landing of Julius Cæsar--he finds himself in a moment not looking at any picture, but standing on the shore among the legionaries, with the whole scene being enacted around him, precisely in every respect as he would have seen it if he had stood there in the flesh on that autumn morning in the year 55 B.C. Since what he sees is but a reflection, the actors are of course entirely unconscious of him, nor can any effort of his change the course of their action in the smallest degree, except only that he can control the rate at which the drama shall pass before him--can have the events of a whole year rehearsed before his eyes in a single hour, or can at any moment stop the movement altogether, and hold any particular scene in view as a picture as long as he chooses.

In truth he observes not only what he would have seen if he had been there at the time in the flesh, but much more. He hears and understands all that the people say, and he is conscious of all their thoughts and motives; and one of the most interesting of the many possibilities which open up before one who has learnt to read the records is the study of the thought of ages long past--the thought of the cave-men and the lake-dwellers as well as that which ruled the mighty civilizations of Atlantis, of Egypt or Chaldaea. What splendid possibilities open up before the man who is in full possession of this power may easily be imagined. He has before him a field of historical research of most entrancing interest. Not only can he review at his leisure all history with which we are acquainted, correcting as he examines it the many errors and misconceptions which have crept into the accounts handed down to us; he can also range at will over the whole story of the world from its very beginning, watching the slow development of intellect in man, the descent of the Lords of the Flame, and the growth of the mighty civilizations which they founded.

Nor is his study confined to the progress of humanity alone; he has before him, as in a museum, all the strange animal and vegetable forms which occupied the stage in days when the world was young; he can follow all the wonderful geological changes which have taken place, and watch the course of the great cataclysms which have altered the whole face of the earth again and again.

In one especial case an even closer sympathy with the past is possible to the reader of the records. If in the course of his enquiries he has to look upon some scene in which he himself has in a former birth taken part, he may deal with it in two ways; he can either regard it in the usual manner as a spectator (though always, be it remembered, as a spectator whose insight and sympathy are perfect) or he may once more identify himself with that long-dead personality of his--may throw himself back for the time into that life of long ago, and absolutely experience over again the thoughts and the emotions, the pleasures and the pains of a prehistoric past. No wilder and more vivid adventures can be conceived than some of those through which he thus may pass; yet through it all he must never lose hold of the consciousness of his own individuality--must retain the power to return at will to his present personality.

It is often asked how it is possible for an investigator accurately to determine the date of any picture from the far-distant past which he disinters from the records. The fact is that it is sometimes rather tedious work to find an exact date, but the thing can usually be done if it is worth while to spend the time and trouble over it. If we are dealing with Greek or Roman times the simplest method is usually to look into the mind of the most intelligent person present in the picture, and see what date he supposes it to be; or the investigator might watch him writing a letter or other document and observe what date, if any, was included in what was written. When once the Roman or Greek date is thus obtained, to reduce it to our own system of chronology is merely a matter of calculation.

Another way which is frequently adopted is to turn from the scene under examination to a contemporary picture in some great and well-known city such as Rome, and note what monarch is reigning there, or who are the consuls for the year; and when such data are discovered a glance at any good history will give the rest. Sometimes a date can be obtained by examining some public proclamation or some legal document; in fact in the times of which we are speaking the difficulty is easily surmounted.

The matter is by no means so simple, however, when we come to deal with periods much earlier than this--with a scene from early Egypt, Chaldæa, or China, or to go further back still, from Atlantis itself or any of its numerous colonies. A date can still be obtained easily enough from the mind of any educated man, but there is no longer any means of relating it to our own system of dates, since the man will be reckoning by eras of which we know nothing, or by the reigns of kings whose history is lost in the night of time.

Our methods, nevertheless, are not yet exhausted. It must be remembered that it is possible for the investigator to pass the records before him at any speed that he may desire--at the rate of a year in a second if he will, or even very much faster still. Now there are one or two events in ancient history whose dates have already been accurately fixed--as, for example, the sinking of Poseidonis in the year 9564 B.C. It is therefore obvious that if from the general appearance of the surroundings it seems probable that a picture seen is within measurable distance of one of these events, it can be related to that event by the simple process of running through the record rapidly, and counting the years between the two as they pass.

Still, if those years ran into thousands, as they might sometimes do, this plan would be insufferably tedious. In that case we are driven back upon the astronomical method. In consequence of the movement which is commonly called the precession of the equinoxes, though it might more accurately be described as a kind of second rotation of the earth, the angle between the equator and the ecliptic steadily but very slowly varies. Thus, after long intervals of time we find the pole of the earth no longer pointing towards the same spot in the apparent sphere of the heavens, or in other words, our pole-star is not, as at present, [Greek: a] Ursæ Minoris, but some other celestial body; and from this position of the pole of the earth, which can easily be ascertained by careful observation of the night-sky of the picture under consideration, an approximate date can be calculated without difficulty.

In estimating the date of occurrences which took place millions of years ago in earlier races, the period of a secondary rotation (or the precession of the equinoxes) is frequently used as a unit, but of course absolute accuracy is not usually required in such cases, round numbers being sufficient for all practical purposes in dealing with epochs so remote.

The accurate reading of the records, whether of one's own past lives or those of others, must not, however, be thought of as an achievement possible to anyone without careful previous training. As has been already remarked, though occasional reflections may be had upon the astral plane, the power to use the mental sense is necessary before any reliable reading can be done. Indeed, to minimize the possibility of error, that sense ought to be fully at the command of the investigator while awake in the physical body; and to acquire that faculty needs years of ceaseless labor and rigid self-discipline.

Many people seem to expect that as soon as they have signed their application and joined the Theosophical Society they will at once remember at least three or four of their past births; indeed, some of them promptly begin to imagine recollections and declare that in their last incarnation they were Mary Queen of Scots, Cleopatra, or Julius Cæsar! Of course such extravagant claims simply bring discredit upon those who are so foolish as to make them but unfortunately some of that discredit is liable to be reflected, however unjustly, upon the Society to which they belong, so that a man who feels seething within him the conviction that he was Homer or Shakespeare would do well to pause and apply common-sense tests on the physical plane before publishing the news to the world.

It is quite true that some people have had glimpses of scenes from their past lives in dreams, but naturally these are usually fragmentary and unreliable. I had myself in earlier life an experience of this nature. Among my dreams I found that one was constantly recurring--a dream of a house with a portico over-looking a beautiful bay, not far from a hill on the top of which rose a graceful building. I knew that house perfectly, and was as familiar with the position of its rooms and the view from its door as I was with those of my home, in this present life. In those days I knew nothing about reincarnation, so that it seemed to me simply a curious coincidence that this dream should repeat itself so often; and it was not until some time after I had joined the Society that, when one who knew was showing me some pictures of my last incarnation, I discovered that this persistent dream had been in reality a partial recollection, and that the house which I knew so well was the one in which I was born more than two thousand years ago.

But although there are several cases on record in which some well-remembered scene has thus come through from one life to another, a considerable development of occult faculty is necessary before an investigator can definitely trace a line of incarnations, whether they be his own or another man's. This will be obvious if we remember the conditions of the problem which has to be worked out. To follow a person from this life to the one preceding it, it is necessary first of all to trace his present life backwards to his birth and then to follow up in reverse order the stages by which the Ego descended into incarnation.

This will obviously take us back eventually to the condition of the Ego upon the higher levels of the mental plane; so it will be seen that to perform this task effectually the investigator must be able to use the sense corresponding to that exalted level while awake in his physical body--in other words, his consciousness must be centred in the reincarnating Ego itself, and no longer in the lower personality. In that case, the memory of the Ego being aroused, his own past incarnations will be spread out before him like an open book, and he would be able, if he wished, to examine the conditions of another Ego upon that level and trace him backwards through the lower mental and astral lives which led up to it, until he came to the last physical death of that Ego, and through it to his previous life.

There is no way but this in which the chain of lives can be followed through with absolute certainty: and consequently we may at once put aside as conscious or unconscious impostors those people who advertise that they are able to trace out anyone's past incarnations for so many shillings a head. Needless to say, the true occultist does not advertise, and never under any circumstances accepts money for any exhibition of his powers.

Assuredly the student who wishes to acquire the power of following up a line of incarnations can do so only by learning from a qualified teacher how the work is to be done. There have been those who persistently asserted that it was only necessary for a man to feel good and devotional and "brotherly," and all the wisdom of the ages would immediately flow in upon him; but a little common-sense will at once expose the absurdity of such a position. However good a child may be, if he wants to know the multiplication table he must set to work and learn it; and the case is precisely similar with the capacity to use spiritual faculties. The faculties themselves will no doubt manifest as the man evolves, but he can learn how to use them reliably and to the best advantage only by steady hard work and persevering effort.

Take the case of those who wish to help others while on the astral plane during sleep; it is obvious that the more knowledge they possess here, the more valuable will their services be on that higher plane. For example, the knowledge of languages would be useful to them, for though on the mental plane men can communicate directly by thought-transference, whatever their languages may be, on the astral plane this is not so, and a thought must be definitely formulated in words before it is comprehensible. If, therefore, you wish to help a man on that plane, you must have some language in common by means of which you can communicate with him, and consequently the more languages you know the more widely useful you will be. In fact there is perhaps no kind of knowledge for which a use cannot be found in the work of the occultist.

It would be well for all students to bear in mind that occultism is the apotheosis of common-sense, and that every vision which comes to them is not necessarily a picture from the âkâshic records, nor every experience a revelation from on high. It is better far to err on the side of healthy scepticism than of over-credulity; and it is an admirable rule never to hunt about for an occult explanation of anything when a plain and obvious physical one is available. Our duty is to endeavour to keep our balance always, and never to lose our self-control, but to take a reasonable, common-sense view of whatever may happen to us; so shall we be better Theosophists, wiser occultists, and more useful helpers than we have ever been before.

As usual, we find examples of all degrees of the power to see into this memory of nature, from the trained man who can consult the record for himself at will, down to the person who gets nothing but occasional vague glimpses, or has even perhaps had only one such glimpse. But even the man who possesses this faculty only partially and occasionally still finds it of the deepest interest. The psychometer, who needs an object physically connected with the past in order to bring it all into life again around him, and the crystal-gazer who can sometimes direct his less certain astral telescope to some historic scene of long ago, may both derive the greatest enjoyment from the exercise of their respective gifts, even though they may not always understand exactly how their results are produced, and may not have them fully under control under all circumstances.

In many cases of the lower manifestations of these powers we find that they are exercised unconsciously; many a crystal-gazer watches scenes from the past without being able to distinguish them from visions of the present, and many a vaguely-psychic person finds pictures constantly arising before his eyes without ever realizing that he is in effect psychometrizing the various objects around him as he happens to touch them or stand near them.

An interesting variant of this class of psychics is the man who is able to psychometrize persons only, and not inanimate objects as is more usual. In most cases this faculty shows itself erratically, so that such a psychic will, when introduced to a stranger, often see in a flash some prominent event in that stranger's earlier life, but on other similar occasions will receive no special impression. More rarely we meet with someone who gets detailed visions of the past life of everyone whom he encounters. Perhaps one of the best

81

examples of this class was the German writer Zschokke, who describes in his autobiography this extraordinary power of which he found himself possessed. He says:--

"It has happened to me occasionally at the first meeting with a total stranger, when I have been listening in silence to his conversation, that his past life up to the present moment, with many minute circumstances belonging to one or other particular scene in it, has come across me like a dream, but distinctly, entirely involuntarily and unsought, occupying in duration a few minutes.

"For a long time I was disposed to consider these fleeting visions as a trick of the fancy--the more so as my dream-vision displayed to me the dress and movements of the actors, the appearance of the room, the furniture, and other accidents of the scene; till on one occasion, in a gamesome mood, I narrated to my family the secret history of a sempstress who had just before quitted the room. I had never seen the person before. Nevertheless the hearers were astonished, and laughed and would not be persuaded but that I had a previous acquaintance with the former life of the person, inasmuch as what I had stated was perfectly true.

"I was not less astonished to find that my dream-vision agreed with reality. I then gave more attention to the subject, and as often as propriety allowed of it, I related to those whose lives had so passed before me the substance of my dream-vision, to obtain from them its contradiction or confirmation. On every occasion its confirmation followed, not without amazement on the part of those who gave it.

"On a certain fair-day I went into the town of Waldshut accompanied by two young foresters, who are still alive. It was evening, and, tired with our walk, we went into an inn called the 'Vine.' We took our supper with a numerous company at the public table, when it happened that they made themselves merry over the peculiarities and simplicity of the Swiss in connection with the belief in mesmerism, Lavater's physiognomical system and the like. One of my companions, whose national pride was touched by their raillery, begged me to make some reply, particularly in answer to a young man of superior appearance who sat opposite, and had indulged in unrestrained ridicule.

"It happened that the events of this person's life had just previously passed before my mind. I turned to him with the question whether he would reply to me with truth and candor if I narrated to him the most secret passages of his history, he being as little known to me as I to him? That would, I suggested, go something beyond Lavater's physiognomical skill. He promised if I told the truth to admit it openly. Then I narrated the events with which my dream-vision had furnished me, and the table learnt the history of the young tradesman's life, of his school years, his peccadilloes, and, finally, of a little act of roguery committed by him on the strong-box of his employer. I described the uninhabited room with its white walls, where to the right of the brown door there had stood upon the table the small black money-chest, etc. The man, much struck, admitted the correctness of each circumstance--even, which I could not expect, of the last."

And after narrating this incident, the worthy Zschokke calmly goes on to wonder whether perhaps after all this remarkable power, which he had so often displayed, might not really have been always the result of mere chance coincidence!

Comparatively few accounts of persons possessing this faculty of looking back into the past are to be found in the literature of the subject, and it might therefore be supposed to be much less common than prevision. I suspect, however, that the truth is rather that it is much less commonly recognized. As I said before, it may very easily happen that a person may see a picture of the past without recognizing it as such, unless there happens to be in it something which attracts special attention, such as a figure in armour or in antique costume. A prevision also might not always be recognized as such at the time; but the occurrence of the event foreseen recalls it vividly at the same time that it manifests its nature, so that it is unlikely to be overlooked. It is probable, therefore, that occasional glimpses of these astral reflections of the âkâshic records are commoner than the published accounts would lead us to believe.

CHAPTER 8

CLAIRVOYANCE IN TIME: THE FUTURE

Even if, in a dim sort of way, we feel ourselves able to grasp the idea that the whole of the past may be simultaneously and actively present in a sufficiently exalted consciousness, we are confronted by a far greater difficulty when we endeavour to realize how all the future may also be comprehended in that consciousness. If we could believe in the Mohammedan doctrine of kismet, or the Calvinistic theory of predestination, the conception would be easy enough, but knowing as we do that both these are grotesque distortions of the truth, we must look round for a more acceptable hypothesis.

There may still be some people who deny the possibility of prevision, but such denial simply shows their ignorance of the evidence on the subject. The large number of authenticated cases leaves no room for doubt as to the fact, but many of them are of such a nature as to render a reasonable explanation by no means easy to find. It is evident that the Ego possesses a certain amount of previsional faculty, and if the events foreseen were always of great importance, one might suppose that an extraordinary stimulus had enabled him for that occasion only to make a clear impression of what he saw upon his lower personality. No doubt that is the explanation of many of the cases in which death or grave disaster is foreseen, but there are a large number of instances on record to which it does not seem to apply, since the events foretold are frequently exceedingly trivial and unimportant.

A well-known story of second-sight in Scotland will illustrate what I mean. A man who had no belief in the occult was forewarned by a Highland seer of the approaching death of a neighbour. The prophecy was given with considerable wealth of detail, including a full description of the funeral, with the names of the four pall-bearers and others who would be present. The auditor seems to have laughed at the whole story and promptly forgotten it, but the death of his neighbor at the time foretold recalled the warning to his mind, and he determined to falsify part of the prediction at any rate by being one of the pall-bearers himself. He succeeded in getting matters arranged as he wished, but just as the funeral was about to start he was called away from his post by some small matter which detained him only a minute or two. As he came hurrying back he saw with surprise that the procession had started without him, and that the prediction had been exactly fulfilled, for the four pall-bearers were those who had been indicated in the vision.

Now here is a very trifling matter, which could have been of no possible importance to anybody, definitely foreseen months beforehand; and although a man makes a determined effort to alter the arrangement indicated he fails entirely to affect it in the least. Certainly this looks very much like predestination, even down to the smallest detail, and it is only when we examine this question from higher planes that we are able to see our way to escape that theory. Of course, as I said before about another branch of the subject, a full explanation eludes us as yet, and obviously must do so until our knowledge is infinitely greater than it is now; the most that we can hope to do for the present is to indicate the line along which an explanation may be found.

There is no doubt whatever that, just as what is happening now is the result of causes set in motion in the past, so what will happen in the future will be the result of causes already in operation. Even down here we can calculate that if certain actions are performed certain results will follow, but our reckoning is constantly liable to be disturbed by the interference of factors which we have not been able to take into account. But if we raise our consciousness to the mental plane we can see very much farther into the results of our actions.

We can trace, for example, the effect of a casual word, not only upon the person to whom it was addressed, but through him on many others as it is passed on in widening circles, until it seems to have affected the whole country; and one glimpse of such a vision is far more efficient than any number of moral precepts in impressing upon us the necessity of extreme circumspection in thought, word, and deed. Not only can we from that plane see thus fully the result of every action, but we can also see where and in what way the results of other actions apparently quite unconnected with it will interfere with and modify it. In fact, it may be said that the results of all causes at present in action are clearly visible--that the future, as it would be if no entirely new causes should arise, lies open before our gaze.

New causes of course do arise, because man's will is free; but in the case of all ordinary people the use which they will make of their freedom can be calculated beforehand with considerable accuracy. The average man has so little real will that he is very much the creature of circumstances; his action in previous lives places him amid certain surroundings, and their influence upon him is so very much the most important factor in his life-story that his future course may be predicted with almost mathematical certainty. With the developed man the case is different; for him also the main events of life are arranged by his past actions, but the way in which he will allow them to affect him, the methods by which he will deal with them and perhaps triumph over them-- these are all his own, and they cannot be foreseen even on the mental plane except as probabilities.

Looking down on man's life in this way from above, it seems as though his free will could be exercised only at certain crises in his career. He arrives at a point in his life where there are obviously two or three alternative courses open before him; he is absolutely free to choose which of them he pleases, and although some one who knew his nature thoroughly well might feel almost certain what his choice would be, such knowledge on his friend's part is in no sense a compelling force.

But when he *has* chosen, he has to go through with it and take the consequences; having entered upon a particular path he may, in many cases, be forced to go on for a very long way before he has any opportunity to turn aside. His position is somewhat like that of the driver of a train; when he comes to a junction he may have the points set either this way or that, and so can pass on to whichever line he pleases, but when he *has* passed on to one of them he is compelled to run on along the line which he has selected until he reaches another set of points, where again an opportunity of choice is offered to him.

Now, in looking down from the mental plane, these points of new departure would be clearly visible, and all the results of each choice would lie open before us, certain to be worked out even to the smallest detail. The only point which would remain uncertain would be the all-important one as to which choice the man would make. We should, in fact, have not one but several futures mapped out before our eyes, without necessarily being able to determine which of them would materialize itself into accomplished fact. In most instances we should see so strong a probability that we should not hesitate to come to a decision, but the case which I have described is certainly theoretically possible. Still, even this much knowledge would enable us to do with safety a good deal of prediction; and it is not difficult for us to imagine that a far higher power than ours might always be able to foresee which way every choice would go, and consequently to prophesy with absolute certainty.

On the buddhic plane, however, no such elaborate process of conscious calculation is necessary, for, as I said before, in some manner which down here is totally inexplicable, the past, the present, and the future, are there all existing simultaneously. One can only accept this fact, for its cause lies in the faculty of the plane, and the way in which this higher faculty works is naturally quite incomprehensible to the physical brain. Yet now and then one may meet with a hint that seems to bring us a trifle nearer to a dim possibility of comprehension. One such hint was given by Dr. Oliver Lodge in his address to the British Association at Cardiff. He said:

"A luminous and helpful idea is that time is but a relative mode of regarding things; we progress through phenomena at a certain definite pace, and this subjective advance we interpret in an objective manner, as if events moved necessarily in this order and at this precise rate. But that may be only one mode of regarding them. The events may be in some sense in existence always, both past and future, and it may be we who are arriving at them, not they which are happening. The analogy of a traveller in a railway train is useful; if he could never leave the train nor alter its pace he would probably consider the landscapes as necessarily successive and be unable to conceive their co-existence.... We perceive, therefore, a possible fourth dimensional aspect about time, the inexorableness of whose flow may be a natural part or our present limitations. And if we once grasp the idea that past and future may be actually existing, we can recognize that they may have a controlling influence on all present action, and the two together may constitute the 'higher plane' or totality of things after which, as it seems to me, we are impelled to seek, in connection with the directing of form or determinism, and the action of living beings consciously directed to a definite and preconceived end."

Time is not in reality the fourth dimension at all; yet to look at it for the moment from that point of view is some slight help towards grasping the ungraspable. Suppose that we hold a wooden cone at right angles to a sheet of paper, and slowly push it through it point first. A microbe living on the surface of that sheet of paper, and having no power of conceiving anything outside of that surface, could not only never see the cone as a whole, but he could form no sort of conception of such a body at all. All that he would see would be the sudden appearance of a tiny circle, which would gradually and mysteriously grow larger and larger until it vanished from his world as suddenly and incomprehensibly as it had come into it.

Thus, what were in reality a series of sections of the cone would appear to him to be successive stages in the life of a circle, and it would be impossible for him to grasp the idea that these successive stages could be seen simultaneously. Yet it is, of course, easy enough for us, looking down upon the transaction from another dimension, to see that the microbe is simply under a delusion arising from its own limitations, and that the cone exists as a whole all the while. Our own delusion as to past, present, and future is possibly not dissimilar, and the view that is gained of any sequence of events from the buddhic plane corresponds to the view of the cone as a whole. Naturally, any attempt to work out this suggestion lands us in a series of startling paradoxes; but the fact remains a fact, nevertheless, and the time will come when it will be clear as noonday to our comprehension.

When the pupil's consciousness is fully developed upon the buddhic plane, therefore, perfect prevision is possible to him, though he may not--nay, he certainly will not--be able to bring the whole result of his sight through fully and in order into this light. Still, a great deal of clear foresight is obviously within his power whenever he likes to exercise it; and even when he is not exercising it, frequent flashes of fore-knowledge come through into his ordinary life, so that he often has an instantaneous intuition as to how things will turn out even before their inception.

Short of this perfect prevision we find, as in the previous cases, that all degrees of this type of clairvoyance exist, from the occasional vague premonitions which cannot in any true sense be called sight at all, up to frequent and fairly complete second-sight. The faculty to which this latter somewhat misleading name has been given is an extremely interesting one, and would well repay more careful and systematic study than has ever hitherto been given to it.

It is best known to us as a not infrequent possession of the Scottish Highlanders, though it is by no means confined to them. Occasional instances of it have appeared in almost every nation, but it has always been commonest among mountaineers and men of lonely life. With us in England it is often spoken of as though it were the exclusive appanage of the Celtic race, but in reality it has appeared among similarly situated peoples the world over. It is stated, for example, to be very common among the Westphalian peasantry.

Sometimes the second-sight consists of a picture clearly foreshowing some coming event; more frequently, perhaps, the glimpse of the future is given by some symbolical appearance. It is noteworthy that the events foreseen are invariably unpleasant ones--death being the commonest of all; I do not recollect a single instance in which the second-sight has shown anything which was not of the most gloomy nature. It has a ghastly symbolism which is all its own--a symbolism of shrouds and corpse-candles, and other funereal horrors. In some cases it appears to be to a certain extent dependent on locality, for it is stated that inhabitants of the Isle of Skye who possess the faculty often lose it when they leave the island, even though it be only to cross to the mainland. The gift of such sight is sometimes hereditary in a family for generations, but this is not an invariable rule, for it often appears sporadically in one member of a family otherwise free from its lugubrious influence.

An example in which an accurate vision of a coming event was seen some months beforehand by second-sight has already been given. Here is another and perhaps a more striking one, which I give exactly as it was related to me by one of the actors in the scene.

"We plunged into the jungle, and had walked on for about an hour without much success, when Cameron, who happened to be next to me, stopped suddenly, turned pale as death, and, pointing straight before him, cried in accents of horror:

"'See! see! merciful heaven, look there!'

"'Where? what? what is it?' we all shouted confusedly, as we rushed up to him and looked round in expectation of encountering a tiger--a cobra--we hardly knew what, but assuredly something terrible, since it had been sufficient to cause such evident emotion in our usually self-contained comrade. But neither tiger nor cobra was visible--nothing but Cameron pointing with ghastly, haggard face and starting eyeballs at something we could not see.

"'Cameron! Cameron' cried I, seizing his arm, "'for heaven's sake, speak! What is the matter?'

"Scarcely were the words out of my mouth when a low, but very peculiar sound struck on my ear, and Cameron, dropping his pointing hand, said in a hoarse, strained voice, 'There! You heard it? Thank God it's over' and fell to the ground insensible.

"There was a momentary confusion while we unfastened his collar, and I dashed in his face some water which I fortunately had in my flask, while another tried to pour brandy between his clenched teeth; and under cover of it I whispered to the man next to me (one of our greatest skeptics, by the way), 'Beauchamp, did *you* hear anything?'

"'Why, yes,' he replied, a curious sound, very; a sort of crash or rattle far away in the distance, yet very distinct; if the thing were not utterly impossible, I could have sworn it was the rattle of musketry.'

"'Just my impression,' murmured I; 'but hush! he is recovering.'

"In a minute or two he was able to speak feebly, and began to thank us and apologize for giving trouble; and soon he sat up, leaning against a tree, and in a firm, though still low voice said:

"'My dear friends, I feel I owe you an explanation of my extraordinary behavior. It is an explanation that I would fain avoid giving; but it must come some time, and so may as well be given now. You may perhaps have noticed that when during our voyage you all joined in scoffing at dreams, portents and visions, I invariably avoided giving any opinion on the subject. I did so because, while I had no desire to court ridicule or provoke discussion, I was unable to agree with you, knowing only too well from my own dread experience that the world which men agree to call that of the supernatural is just as real as--nay, perhaps, even far more real than--this world we see about us. In other words, I, like many of my countrymen, am cursed with the gift of second-sight--that awful faculty which foretells in vision calamities that are shortly to occur.

"'Such a vision I had just now, and its exceptional horror moved me as you have seen. I saw before me a corpse--not that of one who has died a peaceful natural death, but that of the victim of some terrible accident; a ghastly, shapeless mass, with a face swollen, crushed, unrecognizable. I saw this dreadful object placed in a coffin, and the funeral service performed over it. I saw the burial-ground, I saw the clergyman: and though I had never seen either before, I can picture both perfectly in my mind's eye now; I saw you, myself, Beauchamp, all of us and many more, standing round as mourners; I saw the soldiers raise their muskets after the service was over; I heard the volley they fired--and then I knew no more.'

"As he spoke of that volley of musketry I glanced across with a shudder at Beauchamp, and the look of stony horror on that handsome sceptic's face was not to be forgotten."

This is only one incident (and by no means the principal one) in a very remarkable story of psychic experience, but as for the moment we are concerned merely with the example of second-sight which it gives us, I need only say that later in the day the party of young soldiers discovered the body of their commanding officer in the terrible condition so graphically described by Mr. Cameron. The narrative continues:

"When, on the following evening, we arrived at our destination, and our melancholy deposition had been taken down by the proper authorities, Cameron and I went out for a quiet walk, to endeavor with the assistance of the soothing influence of nature to shake off something of the gloom which paralyzed our spirits. Suddenly he clutched my arm, and, pointing through some rude railings, said in a trembling voice, 'Yes, there it is! that is the burial-ground I saw yesterday.' And when later on we were introduced to the chaplain of the post, I noticed, though my friends did not, the irrepressible shudder with which Cameron took his hand, and I knew that he had recognized the clergyman of his vision."

As for the occult rationale of all this, I presume Mr. Cameron's vision was a pure case of second-sight, and if so the fact that the two men who were evidently nearest to him (certainly one--probably both--actually touching him) participated in it to the limited extent of hearing the concluding volley, while the others who were not so close did not, would show that the intensity with which the vision impressed itself upon the seer occasioned vibrations in his mind-body which were communicated to those of the persons in contact with him, as in ordinary thought-transference. Anyone who wishes to read the rest of the story will find it in the pages of *Lucifer*, vol. xx., p. 457.

Scores of examples of similar nature to these might easily be collected. With regard to the symbolical variety of this sight, it is commonly stated among those who possess it that if on meeting a living person they see a phantom shroud wrapped around him, it is a sure prognostication of his death. The date of the approaching decease is indicated either by the extent to which the shroud covers the body, or by the time of day at which the vision is seen; for if it be in the early morning they say that the man will die during the same day, but if it be in the evening, then it will be only some time within a year.

Another variant (and a remarkable one) of the symbolic form of second-sight is that in which the headless apparition of the person whose death is foretold manifests itself to the seer. An example of that class is given in *Signs before Death* as having happened in the family of Dr. Ferrier, though in that case, if I recollect rightly, the vision did not occur until the time of the death, or very near it.

Turning from seers who are regularly in possession of a certain faculty, although its manifestations are only occasionally fully under their control, we are confronted by a large number of isolated instances of prevision in the case of people with whom it is not in any way a regular faculty. Perhaps the majority of these occur in dreams, although examples of the waking vision are by no means wanting. Sometimes the prevision refers to an event of distinct importance to the seer, and so justifies the action of the Ego in taking the trouble to impress it. In other cases, the event is one which is of no apparent importance, or is not in any way connected with the man to whom the vision comes. Sometimes it is clear that the intention of the Ego (or the communicating entity, whatever it may be) is to warn the lower self of the approach of some calamity, either in order that it may be prevented or, if that be not possible, that the shock may be minimized by preparation.

The event most frequently thus foreshadowed is, perhaps not unnaturally, death--sometimes the death of the seer himself, sometimes that of one dear to him. This type of prevision is so common in the literature of the subject, and its object is so obvious, that we need hardly cite examples of it; but one or two instances in which the prophetic sight, though clearly useful, was yet of a less somber character, will prove not uninteresting to the reader. The following is culled from that storehouse of the student of the uncanny, Mrs. Crowe's *Night Side of Nature*, p. 72.

"A few years ago Dr. Watson, now residing at Glasgow, dreamt that he received a summons to attend a patient at a place some miles from where he was living; that he started on horseback, and that as he was crossing a moor he saw a bull making furiously at him, whose horns he only escaped by taking refuge on a spot inaccessible to the animal, where he waited a long time till some people, observing his situation, came to his assistance and released him.

"Whilst at breakfast on the following morning the summons came, and smiling at the odd coincidence (as he thought it), he started on horseback. He was quite ignorant of the road he had to go, but by and by he arrived at the moor, which he recognized, and presently the bull appeared, coming full tilt towards him. But his dream had shown him the place of refuge, for which he instantly made, and there he spent three or four hours, besieged by the animal, till the country people set him free. Dr. Watson declares that but for the dream he should not have known in what direction to run for safety."

Another case, in which a much longer interval separated the warning and its fulfillment, is given by Dr. F. G. Lee, in *Glimpses of the Supernatural*, vol. i., p. 240.

"Mrs. Hannah Green, the housekeeper of a country family in Oxfordshire, dreamt one night that she had been left alone in the house upon a Sunday evening, and that hearing a knock at the door of the chief entrance she went to it and there found an ill-looking tramp armed with a bludgeon, who insisted on forcing himself into the house. She thought that she struggled for some time to prevent him so doing, but quite ineffectually, and that, being struck down by him and rendered insensible, he thereupon gained ingress to the mansion. On this she awoke.

"As nothing happened for a considerable period the circumstance of the dream was soon forgotten, and, as she herself asserts, had altogether passed away from her mind. However, seven years afterwards this same housekeeper was left with two other servants to take charge of an isolated mansion at Kensington (subsequently the town residence of the family), when on a certain Sunday evening, her fellow-servants having gone out and left her alone, she was suddenly startled by a loud knock at the front door.

"All of a sudden the remembrance of her former dream returned to her with singular vividness and remarkable force, and she felt her lonely isolation greatly. Accordingly, having at once lighted a lamp on the hall table--during which act the loud knock was repeated with vigor--she took the precaution to go up to a landing on the stair and throw up the window; and there to her intense terror she saw in the flesh the very man whom years previously she had seen in her dream, armed with the bludgeon and demanding an entrance.

"With great presence of mind she went down to the chief entrance, made that and other doors and windows more secure, and then rang the various bells of the house violently, and placed lights in the upper rooms. It was concluded that by these acts the intruder was scared away."

Evidently in this case also the dream was of practical use, as without it the worthy housekeeper would without doubt from sheer force of habit have opened the door in the ordinary way in answer to the knock.

It is not, however, only in dream that the Ego impresses his lower self with what he thinks it well for it to know. Many instances showing this might be taken from the books, but instead of quoting from them I will give a case related only a few weeks ago by a lady of my acquaintance--a case which, although not surrounded with any romantic incident, has at least the merit of being new.

My friend, then, has two quite young children, and a little while ago the elder of them caught (as was supposed) a bad cold, and suffered for some days from a complete stoppage in the upper part of the nose. The mother thought little of this, expecting it to pass off, until one day she suddenly saw before her in the air what she describes as a picture of a room, in the centre of which was a table on which her child was lying insensible or dead, with some people bending over her. The minutest details of the scene were clear to her, and she particularly noticed that the child wore a white night-dress, whereas she knew that all garments of that description possessed by her little daughter happened to be pink.

This vision impressed her considerably, and suggested to her for the first time that the child might be suffering from something more serious than a cold, so she carried her off to a hospital for examination. The surgeon who attended to her discovered the presence of a dangerous growth in the nose, which he pronounced must be removed. A few days later the child was taken to the hospital for the operation, and was put to bed. When the mother arrived at the hospital she found she had forgotten to bring one of the child's night-dresses, and so the nurses had to supply one, which was *white*. In this white dress the operation was performed on the girl the next day, in the room that her mother saw in her vision, every circumstance being exactly reproduced.

In all these cases the prevision achieved its result, but the books are full of stories of warnings neglected or scouted, and of the disaster that consequently followed. In some cases the information is given to someone who has practically no power to interfere in the matter, as in the historic instance when John Williams, a Cornish mine-manager, foresaw in the minutest detail, eight or nine days before it took place, the assassination of Mr. Spencer Perceval, the then Chancellor of the Exchequer, in the lobby of the House of Commons. Even in this case, however, it is just possible that something might have been done, for we read that Mr. Williams was so much impressed that he consulted his friends as to whether he ought not to go up to London to warn Mr. Perceval. Unfortunately they dissuaded him, and the assassination took place. It does not seem very probable that, even if he had gone up to town and related his story, much attention would have been paid to him, still there is just the possibility that some precautions might have been taken which would have prevented the murder.

There is little to show us what particular action on higher planes led to this curious prophetic vision. The parties were entirely unknown to one another, so that it was not caused by any close sympathy between them. If it was an attempt made by some helper to avert the threatened doom, it seems strange that no one who was sufficiently impressible could be found nearer than Cornwall. Perhaps Mr. Williams, when on the astral plane during sleep, somehow came across this reflection of the future, and being naturally horrified thereby, passed it on to his lower mind in the hope that somehow something might be done to prevent it; but it is impossible to diagnose the case with certainty without examining the âkâshic records to see what actually took place.

A typical instance of the absolutely purposeless foresight is that related by Mr. Stead, in his *Real Ghost Stories* (p. 83), of his friend Miss Freer, commonly known as Miss X. When staying at a country house this lady, being wide awake and fully conscious, once saw a dogcart drawn by a white horse standing at the hall door, with two strangers in it, one of whom got out of the cart and stood playing with a terrier. She noticed that he was wearing an ulster, and also particularly observed the fresh wheel-marks made by the cart on the gravel. Nevertheless there was no cart there at the time; but half an hour later two strangers *did* drive up in such an equipage, and every detail of the lady's vision was accurately fulfilled. Mr. Stead goes on to cite another instance of equally purposeless prevision where seven years separated the dream (for in this case it was a dream) and its fulfillment.

All these instances (and they are merely random selections from many hundreds) show that a certain amount of prevision is undoubtedly possible to the Ego, and such cases would evidently be much more frequent if it were not for the exceeding density and lack of response in the lower vehicles of the majority of what we call civilized mankind--qualities chiefly attributable to the gross practical materialism of the present age. I am not thinking of any profession of materialistic belief as common, but of the fact that in all practical affairs of daily life nearly everyone is guided solely by considerations of worldly interest in some shape or other.

In many cases the Ego himself may be an undeveloped one, and his prevision consequently very vague; in others he himself may see clearly, but may find his lower vehicles so unimpressible that all he can succeed in getting through into his physical brain may be an indefinite presage of coming disaster. Again, there are cases in which a premonition is not the work of the Ego at all, but of some outside entity, who for some reason takes a friendly interest in the person to whom the feeling comes. In the work which I quoted above, Mr. Stead tells us of the certainty which he felt many months beforehand that be would be left in charge of the *Pall Mall Gazette* though from an ordinary point of view nothing seemed less probable. Whether that fore-knowledge was the result of an impression made by his own Ego or of a friendly hint from someone else it is impossible to say without definite investigation, but his confidence in it was fully justified.

There is one more variety of clairvoyance in time which ought not to be left without mention. It is a comparatively rare one, but there are enough examples on record to claim our attention, though unfortunately the particulars given do not usually include those which we should require in order to be able to diagnose it with certainty. I refer to the cases in which spectral armies or phantom flocks of animals have been seen. In *The Night Side of Nature* (p. 462 *et seq.*) we have accounts of several such visions. We are there told how at Havarah Park, near Ripley, a body of soldiers in white uniform, amounting to several hundreds, was seen by reputable people to go through various evolutions and then vanish; and how some years earlier a similar visionary army was seen in the neighborhood of Inverness by a respectable farmer and his son.

In this case also the number of troops was very great, and the spectators had not the slightest doubt at first that they were substantial forms of flesh and blood. They counted at least sixteen pairs of columns, and had abundance of time to observe every particular. The front ranks marched seven abreast, and were accompanied by a good many women and children, who were carrying tin cans and other implements of cookery. The men were clothed in red, and their arms shone brightly in the sun. In the midst of them was an animal, a deer or a horse, they could not distinguish which, that they were driving furiously forward with their bayonets.

The younger of the two men observed to the other that every now and then the rear ranks were obliged to run to overtake the van; and the elder one, who had been a soldier, remarked that that was always the case, and recommended him if he ever served to try to march in the front. There was only one mounted officer; he rode a grey dragoon horse, and wore a gold-laced hat and blue Hussar cloak, with wide open sleeves lined with red. The two spectators observed him so particularly that they said afterwards they should recognize him anywhere. They were, however, afraid of being ill-treated or forced to go along with the troops, whom they concluded to have come from Ireland, and landed at Kyntyre; and whilst they were climbing over a dyke to get out of their way, the whole thing vanished.

A phenomenon of the same sort was observed in the earlier part of this century at Paderborn in Westphalia, and seen by at least thirty people; but as, some years later, a review of twenty thousand men was held on the very same spot, it was concluded that the vision must have been some sort of second-sight--a faculty not uncommon in the district.

Such spectral hosts, however, are sometimes seen where an army of ordinary men could by no possibility have marched, either before or after. One of the most remarkable accounts of such apparitions is given by Miss Harriet Martineau, in her description of *The English Lakes.* She writes as follows:--

"This Souter or Soutra Fell is the mountain on which ghosts appeared in myriads, at intervals during ten years of the last century, presenting the same appearances to twenty-six chosen witnesses, and to all the inhabitants of all the cottages within view of the mountain, and for a space of two hours and a half at one time--the spectral show being closed by darkness! The mountain, be it remembered, is full of precipices, which defy all marching of bodies of men; and the north and west sides present a sheer perpendicular of 900 feet.

"On Midsummer Eve, 1735, a farm servant of Mr. Lancaster, half a mile from the mountain, saw the eastern side of its summit covered with troops, which pursued their onward march for an hour. They came, in distinct bodies, from an eminence on the north end, and disappeared in a niche in the summit. When the poor fellow told his tale, he was insulted on all hands, as original observers usually are when they see anything wonderful. Two years after, also on a Midsummer Eve, Mr. Lancaster saw some men there, apparently following their horses, as if they had returned from hunting. He thought nothing of this; but he happened to look up again ten minutes after, and saw the figures, now mounted, and followed by an interminable array of troops, five abreast, marching from the eminence and over the cleft as before. All the family saw this, and the maneuvers of the force, as each company was kept in order by a mounted officer, who galloped this way and that. As the shades of twilight came on, the discipline appeared to relax, and the troops intermingled, and rode at unequal paces, till all was lost in darkness. Now of course all the Lancasters were insulted, as their servant had been; but their justification was not long delayed.

"On the Midsummer Eve of the fearful 1745, twenty-six persons, expressly summoned by the family, saw all that had been seen before, and more. Carriages were now interspersed with the troops; and everybody knew that no carriages had been, or could be, on the summit of Souter Fell. The multitude was beyond imagination; for the troops filled a space of half a mile, and marched quickly till night hid them--still marching. There was nothing vaporous or indistinct about the appearance of these specters. So real did they seem, that some of the people went up, the next morning, to look for the hoof-marks of the horses; and awful it was to them to find not one foot-print on heather or grass. The witnesses attested the whole story on oath before a magistrate; and fearful were the expectations held by the whole country-side about the coming events of the Scotch rebellion.

"It now comes out that two other persons had seen something of the sort in the interval--*viz.*, in 1743--but had concealed it, to escape the insults to which their neighbors were subjected. Mr. Wren, of Wilton Hall, and his farm servant, saw, one summer evening, a man and a dog on the mountain, pursuing some horses along a place so steep that a horse could hardly by any possibility keep a footing on it. Their speed was prodigious, and their disappearance at the south end of the fell so rapid, that Mr. Wren and the servant went up, the next morning, to find the body of the man who must have been killed. Of man, horse, or dog, they found not a trace and they came down and held their tongues. When they did speak, they fared not much better for having twenty-six sworn comrades in their disgrace.

"As for the explanation, the editor of the *Lonsdale Magazine* declared (vol. ii., p. 313) that it was discovered that on the Midsummer Eve of 1745 the rebels were 'exercising on the western coast of Scotland, whose movements had been reflected by some transparent vapor, similar to the Fata Morgana.' This is not much in the way of explanation; but it is, as far as we know, all that can be had at present. These facts, however, brought out a good many more; as the spectral march of the same kind seen in Leicestershire in 1707, and the tradition of the tramp of armies over Helvellyn, on the eve of the battle of Marston Moor."

Other cases are cited in which flocks of spectral sheep have been seen on certain roads, and there are of course various German stories of phantom cavalcades of hunters and robbers.

Now in these cases, as so often happens in the investigation of occult phenomena, there are several possible causes, any one of which would be quite adequate to the production of the observed occurrences, but in the absence of fuller information it is hardly feasible to do more than guess as to which of these possible causes were in operation in any particular instance.

The explanation usually suggested (whenever the whole story is not ridiculed as a falsehood) is that what is seen is a reflection by mirage of the movements of a real body of troops, taking place at a considerable distance. I have myself seen the ordinary mirage on several occasions, and know something therefore of its wonderful powers of deception; but it seems to me that we should need some entirely new variety of mirage, quite different from that at present known to science, to account for these tales of phantom armies, some of which pass the spectator within a few yards.

First of all, they may be, as apparently in the Westphalian case above mentioned, simply instances of prevision on a gigantic scale- -by whom arranged, and for what purpose, it is not easy to divine. Again, they may often belong to the past instead of the future, and be in fact the reflection of scenes from the âkâshic records-- though here again the reason and method of such reflection is not obvious.

There are plenty of tribes of nature-spirits perfectly capable, if for any reason they wished to do so, of producing such appearances by their wonderful power of glamour (see *Theosophical Manual, No. V.*, p. 60), and such action would be quite in keeping with their delight in mystifying and impressing human beings. Or it may even sometimes be kindly intended by them as a warning to their friends of events that they know to be about to take place. It seems as though some explanation along these lines would be the most reasonable method of accounting for the extraordinary series of phenomena described by Miss Martineau--that is, if the stories told to her can be relied upon.

Another possibility is that in some cases what have been taken for soldiers were simply the nature-spirits themselves going through some of the ordered evolutions in which they take so much delight, though it must be admitted that these are rarely of a character which could be mistaken for military maneuvers except by the most ignorant.

The flocks of animals are probably in most instances mere records, but there are cases where they, like the "wild huntsmen" of German story, belong to an entirely different class of phenomena, which is altogether outside of our present subject. Students of the occult will be familiar with the fact that the circumstances surrounding any scene of intense terror or passion, such as an exceptionally horrible murder, are liable to be occasionally reproduced in a form which it needs a very slight development of psychic faculty to be able to see and it has sometimes happened that various animals formed part of such surroundings, and consequently they also are periodically reproduced by the action of the guilty conscience of the murderer (see *Manual V.*, p. 83).

Probably whatever foundation of fact underlies the various stories of spectral horsemen and hunting-troops may generally be referred to this category. This is also the explanation, evidently, of some of the visions of ghostly armies, such as that remarkable re-enactment of the battle of Edgehill which seems to have taken place at intervals for some months after the date of the real struggle, as testified by a justice of the peace, a clergyman, and other eye-witnesses, in a curious contemporary pamphlet entitled *Prodigious Noises of War and Battle, at Edgehill, near Keinton, in Northamptonshire*. According to the pamphlet this case was investigated at the time by some officers of the army, who clearly recognized many of the phantom figures that they saw. This looks decidedly like an instance of the terrible power of man's unrestrained passions to reproduce themselves, and to cause in some strange way a kind of materialization of their record.

In some cases it is clear that the flocks of animals seen have been simply hordes of unclean artificial elementals taking that form in order to feed upon the loathsome emanations of peculiarly horrible places, such as would be the site of a gallows. An instance of this kind is furnished by the celebrated "Gyb Ghosts," or ghosts of the gibbet, described in *More Glimpses of the World Unseen*, p. 109, as being repeatedly seen in the form of herds of misshapen swine-like creatures, rushing, rooting and fighting night after night on the site of that foul monument of crime. But these belong to the subject of apparitions rather than to that of clairvoyance.

CHAPTER 9

METHODS OF DEVELOPMENT

When a man becomes convinced of the reality of the valuable power of clairvoyance, his first question usually is, "How can I develop in my own case this faculty which is said to be latent in everyone?"

Now the fact is that there are many methods by which it may be developed, but only one which can be at all safely recommended for general use--that of which we shall speak last of all. Among the less advanced nations of the world the clairvoyant state has been produced in various objectionable ways; among some of the non-Aryan tribes of India, by the use of intoxicating drugs or the inhaling of stupefying fumes; among the dervishes, by whirling in a mad dance of religious fervor until vertigo and insensibility supervene; among the followers of the abominable practices of the Voodoo cult, by frightful sacrifices and loathsome rites of black magic. Methods such as these are happily not in vogue in our own race, yet even among us large numbers of dabblers in this ancient art adopt some plan of self-hypnotization, such as the gazing at a bright spot or the repetition of some formula until a condition of semi-stupefaction is produced; while yet another school among them would endeavor to arrive at similar results by the use of some of the Indian systems of regulation of the breath.

All these methods are unequivocally to be condemned as quite unsafe for the practice of the ordinary man who has no idea of what he is doing--who is simply making vague experiments in an unknown world. Even the method of obtaining clairvoyance by allowing oneself to be mesmerized by another person is one from which I should myself shrink with the most decided distaste; and assuredly it should never be attempted except under conditions of absolute trust and affection between the magnetizer and the magnetized, and a perfection of purity in heart and soul, in mind and intention, such as is rarely to be seen among any but the greatest of saints.

Experiments in connection with the mesmeric trance are of the deepest interest, as offering (among other things) a possibility of proof of the fact of clairvoyance to the skeptic, yet except under such conditions as I have just mentioned--conditions, I quite admit, almost impossible to realize--I should never counsel anyone to submit himself as a subject for them.

Curative mesmerism (in which, without putting the patient into the trance state at all, an effort is made to relieve his pain, to remove his disease, or to pour vitality into him by magnetic passes) stands on an entirely different footing; and if the mesmerizer, even though quite untrained, is himself in good health and animated by pure intentions, no harm is likely to be done to the subject. In so extreme a case as that of a surgical operation, a man might reasonably submit himself even to the mesmeric trance, but it is certainly not a condition with which one ought lightly to experiment. Indeed, I should most strongly advise any one who did me the honor to ask for my opinion on the subject, not to attempt any kind of experimental investigation into what are still to him the abnormal forces of nature, until he has first of all read carefully everything that has been written on the subject, or-- which is by far the best of all--until he is under the guidance of a qualified teacher.

But where, it will be said, is the qualified teacher to be found? Not, most assuredly, among any who advertise themselves as teachers, who offer to impart for so many guineas or dollars the sacred mysteries of the ages, or hold "developing circles" to which casual applicants are admitted at so much per head.

Much has been said in this treatise of the necessity for careful training--of the immense advantages of the trained over the untrained clairvoyant; but that again brings us back to the same question--where is this definite training to be had?

The answer is, that the training may be had precisely where it has always been to be found since the world's history began--at the hands of the Great White Brotherhood of Adepts, which stands now, as it has always stood, at the back of human evolution, guiding and helping it under the sway of the great cosmic laws which represent to us the Will of the Eternal.

But how, it may be asked, is access to be gained to them? How is the aspirant thirsting for knowledge to signify to them his wish for instruction?

Once more, by the time-honored methods only. There is no new patent whereby a man can qualify himself without trouble to become a pupil in that School--no royal road to the learning which has to be acquired in it. At the present day, just as in the mists of antiquity, the man who wishes to attract their notice must enter upon the slow and toilsome path of self-development--must learn first of all to take himself in hand and make himself all that he ought to be. The steps of that path are no secret; I have given them in full detail in *Invisible Helpers*, so I need not repeat them here. But it is no easy road to follow, and yet sooner or later all must follow it, for the great law of evolution sweeps mankind slowly but resistlessly towards its goal.

From those who are pressing into this path the great Masters select their pupils, and it is only by qualifying himself to be taught that a man can put himself in the way of getting the teaching. Without that qualification, membership in any Lodge or Society, whether secret or otherwise, will not advance his object in the slightest degree. It is true, as we all know, that it was at the instance of some of these Masters that our Theosophical Society was founded, and that from its ranks some have been chosen to pass into closer relations with them. But that choice depends upon the earnestness of the candidate, not upon his mere membership of the Society or of any body within it.

That, then, is the only absolutely safe way of developing clairvoyance--to enter with all one's energy upon the path of moral and mental evolution, at one stage of which this and other of the higher faculties will spontaneously begin to show themselves. Yet there is one practice which is advised by all the religions alike--which if adopted carefully and reverently can do no harm to any human being, yet from which a very pure type of clairvoyance has sometimes been developed; and that is the practice of meditation.

Let a man choose a certain time every day--a time when he can rely upon being quiet and undisturbed, though preferably in the daytime rather than at night--and set himself at that time to keep his mind for a few minutes entirely free from all earthly thoughts of any kind whatever and, when that is achieved, to direct the whole force of his being towards the highest spiritual ideal that he happens to know. He will find that to gain such perfect control of thought is enormously more difficult than he supposes, but when he attains it it cannot but be in every way most beneficial to him, and as he grows more and more able to elevate and concentrate his thought, he may gradually find that new worlds are opening before his sight.

As a preliminary training towards the satisfactory achievement of such meditation, he will find it desirable to make a practice of concentration in the affairs of daily life--even in the smallest of them. If he writes a letter, let him think of nothing else but that letter until it is finished if he reads a book, let him see to it that his thought is never allowed to wander from his author's meaning. He must learn to hold his mind in check, and to be master of that also, as well as of his lower passions he must patiently labor to acquire absolute control of his thoughts, so that he will always know exactly what he is thinking about, and why--so that he can use his mind, and turn it or hold it still, as a practiced swordsman turns his weapon where he will.

Yet after all, if those who so earnestly desire clairvoyance could possess it temporarily for a day or even an hour, it is far from certain that they would choose to retain the gift. True, it opens before them new worlds of study, new powers of usefulness, and for this latter reason most of us feel it worth while; but it should be remembered that for one whose duty still calls him to live in the world it is by no means an unmixed blessing. Upon one in whom that vision is opened the sorrow and the misery, the evil and the greed of the world press as an ever-present burden, until in the earlier days of his knowledge he often feels inclined to echo the

107

passionate adjuration contained in those rolling lines of Schiller's:

Dien Orakel zu verkünden, warum warfest du mich hin In die Stadt der ewig Blinden, mit dem aufgeschloss'nen Sinn? Frommt's, den Schleier aufzuheben, wo das nahe Schreckniss droht? Nur der Irrthum ist das Leben; dieses Wissen ist der Tod. Nimm, O nimm die traur'ge Klarheit mir vom Aug' den blut'gen Schein! Schrecklich ist es deiner Wahrheit sterbliches Gefäss zu seyn!

which may perhaps be translated "Why hast thou cast me thus into the town of the ever-blind, to proclaim thine oracle by the opened sense? What profits it to lift the veil where the near darkness threatens? Only ignorance is life; this knowledge is death. Take back this sad clear-sightedness; take from mine eyes this cruel light! It is horrible to be the mortal channel of thy truth." And again later he cries, "Give me back my blindness, the happy darkness of my senses; take back thy dreadful gift!"

But this of course is a feeling which passes, for the higher sight soon shows the pupil something beyond the sorrow--soon bears in upon his soul the overwhelming certainty that, whatever appearances down here may seem to indicate, all things are without shadow of doubt working together for the eventual good of all. He reflects that the sin and the suffering are there, whether he is able to perceive them or not, and that when he can see them he is after all better able to give efficient help than he would be if he were working in the dark; and so by degrees he learns to bear his share of the heavy karma of the world.

Some misguided mortals there are who, having the good fortune to possess some slight touch of this higher power, are nevertheless so absolutely destitute of all right feeling in connection with it as to use it for the most sordid ends--actually even to advertise themselves as "test and business clairvoyants!" Needless to say, such use of the faculty is a mere prostitution and degradation of it, showing that its unfortunate possessor has somehow got hold of it before the moral side of his nature has been sufficiently developed to stand the strain which it imposes. A perception of the amount of evil karma that may be generated by such action in a very short time changes one's disgust into pity for the unhappy perpetrator of that sacrilegious folly.

It is sometimes objected that the possession of clairvoyance destroys all privacy, and confers a limit-less ability to explore the secrets of others. No doubt it does confer such an *ability*, but nevertheless the suggestion is an amusing one to anyone who knows anything practically about the matter. Such an objection may possibly be well-founded as regards the very limited powers of the "test and business clairvoyant," but the man who brings it forward against those who have had the faculty opened for them in the course of their instruction, and consequently possess it fully, is forgetting three fundamental facts: first, that it is quite inconceivable that anyone, having before him the splendid fields for investigation which true clairvoyance opens up, could ever have the slightest wish to pry into the trumpery little secrets of any individual man; secondly, that even if by some impossible chance our clairvoyant *had* such indecent curiosity about matters of petty gossip, there is, after all, such a thing as the honor of a gentleman, which, on that plane as on this, would of course prevent him from contemplating for an instant the idea of gratifying it; and thirdly, in case, by any unheard-of possibility, one might encounter some variety of low-class pitri with whom the above considerations would have no weight, full instructions are always given to every pupil, as soon as he develops any sign of faculty, as to the limitations which are placed upon its use.

Put briefly, these restrictions are that there shall be no prying, no selfish use of the power, and no displaying of phenomena. That is to say, that the same considerations which would govern the actions of a man of right feeling upon the physical plane are expected to apply upon the astral and mental planes also; that the pupil is never under any circumstances to use the power which his additional knowledge gives to him in order to promote his own worldly advantage, or indeed in connection with gain in any way; and that he is never to give what is called in spiritualistic circles "a test"--that is, to do anything which will incontestably prove to skeptics on the physical plane that he possesses what to them would appear to be an abnormal power.

With regard to this latter proviso people often say, "But why should he not? it would be so easy to confute and convince your skeptic, and it would do him good!" Such critics lose sight of the fact that, in the first place, none of those who know anything *want* to confute or convince skeptics, or trouble themselves in the slightest degree about the skeptic's attitude one way or the other; and in the second, they fail to understand how much better it is for that skeptic that he should gradually grow into an intellectual appreciation of the facts of nature, instead of being suddenly introduced to them by a knock-down blow, as it were. But the subject was fully considered many years ago in Mr. Sinnet's *Occult World*, and it is needless to repeat again the arguments there adduced.

It is very hard for some of our friends to realize that the silly gossip and idle curiosity which so entirely fill the lives of the brainless majority on earth can have no place in the more real life of the disciple; and so they sometimes enquire whether, even without any special wish to see, a clairvoyant might not casually observe some secret which another person was trying to keep, in the same way as one's glance might casually fall upon a sentence in someone else's letter which happened to be lying open upon the table. Of course he might, but what if he did? The man of honor would at once avert his eyes, in one case as in the other, and it would be as though he had not seen. If objectors could but grasp the idea that no pupil *cares* about other people's business, except when it comes within his province to try to help them, and that he has always a world of work of his own to attend to, they would not be so hopelessly far from understanding the facts of the wider life of the trained clairvoyant.

Even from the little that I have said with regard to the restrictions laid upon the pupil, it will be obvious that in very many cases he will know much more than he is at liberty to say. That is of course true in a far wider sense of the great Masters of Wisdom themselves, and that is why those who have the privilege of occasionally entering their presence pay so much respect to their lightest word even on subjects quite apart from the direct teaching. For the opinion of a Master, or even of one of his higher pupils, upon any subject is that of a man whose opportunity of judging accurately is out of all proportion to ours.

His position and his extended faculties are in reality the heritage of all mankind, and, far though we may now be from those grand powers, they will none the less certainly be ours one day. Yet how

110

different a place will this old world be when humanity as a whole possesses the higher clairvoyance! Think what the difference will be to history when all can read the records; to science, when all the processes about which now men theorize can be watched through all their course; to medicine, when doctor and patient alike can see clearly and exactly all that is being done; to philosophy, when there is no longer any possibility of discussion as to its basis, because all alike can see a wider aspect of the truth; to labor, when all work will be joy, because every man will be put only to that which he can do best; to education, when the minds and hearts of the children are open to the teacher who is trying to form their character; to religion, when there is no longer any possibility of dispute as to its broad dogmas, since the truth about the states after death, and the Great Law that governs the world, will be patent to all eyes.

Above all, how far easier it will be then for the evolved men to help one another under those so much freer conditions! The possibilities that open before the mind are as glorious vistas stretching in all directions, so that our seventh round should indeed be a veritable golden age. Well for us that these grand faculties will not be possessed by all humanity until it has evolved to a far higher level in morality as well as in wisdom, else should we but repeat once more under still worse conditions the terrible downfall of the great Atlantean civilization, whose members failed to realize that increased power meant increased responsibility. Yet we ourselves were most of us among those very men let us hope that we have learnt wisdom by that failure, and that when the possibilities of the wider life open before us once more, this time we shall bear the trial better.

THE FOLLOWING WAS INCLUDED IN THE FIRST EDITION

The Theosofical Society is composed of students, belonging to any religion in the world or to none, who are united by their approval of the above objects, by their wish to remove religious antagonisms and to draw together men of good-will whatsoever their religious opinions, and by their desire to study religious truths and to share the results of their studies with others. Their bond of union is not the profession of a common belief, but a common search and aspiration for Truth. They hold that Truth should be sought by study, by reflection, by purity of life, by devotion to high ideals, and they regard Truth as a prize to be striven for, not as a dogma to be imposed by authority. They consider that belief should be the result of individual study or intuition, and not its antecedent, and should rest on knowledge, not on assertion. They extend tolerance to all, even to the intolerant, not as a privilege they bestow, but as a duty they perform, and they seek to remove ignorance, not to punish it. They see every religion as an expression of the DIVINE WISDOM, and prefer its study to its condemnation, and its practice to proselytism. Peace is their watch-word, as Truth is their aim.

THEOSOPHY is the body of truths which forms the basis of all religions, and which cannot be claimed as the exclusive possession of any. It offers a philosophy which renders life intelligible, and which demonstrates the justice and the love which guide its evolution. It puts death in its rightful place, as a recurring incident in an endless life, opening the gateway of a fuller and more radiant existence. It restores to the world the science of the spirit, teaching man to know the spirit as himself, and the mind and body as his servants. It illuminates the scriptures and doctrines of religions by unveiling their hidden meanings, and thus justifying them at the bar of intelligence, as they are ever justified in the eyes of intuition.

Members of the Theosophical Society study these truths, and Theosophists endeavour to live them. Every one willing to study, to be tolerant, to aim high, and to work perseveringly, is welcomed as a member, and it rests with the member to become a true Theosophist.

BOOKS RECOMMENDED FOR STUDY

An Outline of Theosophy. C. W. Leadbeater - Ancient Wisdom. Annie Besant - Theosophical Manuals. Seven Principles of Man. Annie Besant - Re-incarnation. Annie Besant - Karma. Annie Besant - Death--and After? Annie Besant - The Astral Plane. C. W. Leadbeater - The Devachanic Plane. C. W. Leadbeater - Man and his Bodies. Annie Besant - The Key to Theosophy. H. P. Blavatsky - Esoteric Buddhism. A. P. Sinnett - The Growth of the Soul. A. P. Sinnett - Man's Place in the Universe - Man Visible and Invisible (illustrated). C. W. Leadbeater

A student who has thoroughly mastered these may study The Secret Doctrine. H. P. Blavatsky. Three volumes and separate index. Man Visible and Invisible (illustrated). C. W. Leadbeater

WORLD-RELIGIONS. s. d. Fragments of a Faith Forgotten. G. R. S. Mead - Esoteric Christianity. Annie Besant - Four Great Religions. Annie Besant - Orpheus. G. R. S. Mead - The Kabalah. A. E. Waite

ETHICAL. In the Outer Court. Annie Besant - The Path of Discipleship. Annie Besant - The Voice of the Silence. H. P. Blavatsky - Light on the Path. Mabel Collins - Bhagavad-Gitâ. Trans. Annie Besant - Studies in the Bhagavad-Gitâ - The Doctrine of the Heart - The Upanishats. Trans. by G. R. S. Mead and J.C. Chattopadyaya. Two Volumes - Three Paths and Dharma. Annie Besant - Theosophy of the Upanishats - The Stanzas of Dayân. H.P. Blavatsky

VARIOUS. Nature's Mysteries. A. P. Sinnett - Clairvoyance. C. W. Leadbeater - Dreams. C. W. Leadbeater - The Building of the Kosmos. Annie Besant - The Evolution of Life and Form. Annie Besant - Some Problems of Life. Annie Besant - Thought-Power, its Control and Culture. Annie Besant - The Science of the Emotions. Bhagavan Das - The Gospel and the Gospels. G. R. S. Mead - Five Years of Theosophy

CPSIA information can be obtained at www.ICGtesting.com
Printed in the USA
LVOW07s2213271015

460058LV00024B/704/P